# Air Pollution and Freshwater Ecosystems

## Sampling, Analysis, and Quality Assurance

# Air Pollution and Freshwater Ecosystems

## Sampling, Analysis, and Quality Assurance

Timothy J. Sullivan
Alan T. Herlihy
James R. Webb

CRC Press
Taylor & Francis Group
Boca Raton London New York

CRC Press is an imprint of the
Taylor & Francis Group, an **informa** business

Cover Image: Photograph of a Virginia stream taken by J.R. Webb

CRC Press
Taylor & Francis Group
6000 Broken Sound Parkway NW, Suite 300
Boca Raton, FL 33487-2742

© 2015 by Taylor & Francis Group, LLC
CRC Press is an imprint of Taylor & Francis Group, an Informa business

Printed on acid-free paper
Version Date: 20141009

International Standard Book Number-13: 978-1-4822-2713-0 (Hardback)

---

### Library of Congress Cataloging-in-Publication Data

Sullivan, Timothy J., 1950 July 17-
   Air pollution and freshwater ecosystems : sampling, analysis, and quality assurance / Timothy J. Sullivan, Alan T. Herlihy, and James R. Webb.
      pages cm
   Includes bibliographical references and index.
   ISBN 978-1-4822-2713-0
   1. Water quality--Measurement--Technique. 2. Air pollution--Environmental aspects--Measurement--Technique. 3. Freshwater ecology. I. Herlihy, Alan T., 1958- II. Webb, James R., 1948- III. Title.

TD367.S85 2014
628.1'61--dc23
                                                                                      2014021321
---

Visit the Taylor & Francis Web site at
http://www.taylorandfrancis.com

and the CRC Press Web site at
http://www.crcpress.com

# Contents

## Appendix C: Basic Standard Operating Procedures for Lake Field-Sampling Activities

# List of Figures

# List of Tables

# Preface

Air pollution adversely affects lakes and streams (collectively called surface waters) and their watersheds throughout the developed world. Some of the most widely studied air pollutants that affect freshwaters include acid precursors such as sulfur (S) and nitrogen (N); nutrients (mainly N); and toxics such as mercury (Hg), pesticides, and combustion by-products. These pollutants can have an impact on water quality and harm sensitive species of aquatic biota. They also alter the chemistry of watershed soils, with potential impacts on plant roots and other terrestrial life-forms. Scientists study freshwater to improve understanding of processes that govern the biogeochemistry of the entire landscape, not just the aquatic portions of it. The chemistry of drainage water integrates a host of terrestrial and aquatic processes, including acidification, nutrient cycling, and bioaccumulation of toxic substances. These processes interact with water as it moves from the atmosphere as precipitation through the soil and into the groundwater and eventually to streams, lakes, rivers, and estuaries.

To develop an understanding of water quality, scientists collect samples of lake and stream water, measure in a laboratory the concentrations of various chemical constituents, document the quality of the resulting data, and ascertain potential effects on sensitive life-forms that inhabit that water. Specific kinds of data analyses, including statistical applications, are conducted. This book provides instructions regarding how to do all of that. We summarize the collective experience of three researchers who have been studying the effects of air pollutants on soils, waters, and associated biota across the United States for the past three decades. The presentation includes many examples of sampling, analysis, and interpretation techniques. This book is targeted to students and practitioners of biogeochemistry, water resources, and aquatic and terrestrial ecology and to water resource professionals and other scientists and resource managers.

A host of questions revolve around, and depend on, the study of water quality. What lakes are acid sensitive or acid affected? Is a stream limited in its primary productivity by N, phosphorus (P), light, or something else? Have mayflies, zooplankton, or fish been impacted by too much acidity? If so, which species? Do pesticide applications in the lower-elevation agricultural lands affect amphibians in the nearby mountain lakes? What is the critical load of air pollution that sensitive downwind resources can tolerate without unacceptable damage? What level of damage is acceptable? Does a lake watershed receive too much N from the air? Are the water quality conditions getting better, staying the same, or getting worse over time? The questions are endless. To answer them, we need data; not just any data, but good quality-assured data that are properly collected, analyzed, and interpreted. If you have questions such as these or if you want answers that you can use as a basis for policy or management decisions, this book can be of assistance.

There is not just one correct way to collect a lake or stream water sample. Neither is there one way to conduct a laboratory analysis, quality assure a database, analyze and interpret a suite of chemical analysis results, or evaluate the integrity of an aquatic invertebrate community. Therefore, the information presented in this book is not structured as if it was a cookbook. Rather, we provide recommendations for how things might be done; even more important, we explain why we make the recommendations that we make. We alert you, the reader, to the potential problems and issues that we have considered in making decisions about how to proceed. If you understand the reasons why we recommend that certain steps be taken, you will be in a much stronger position to make your own changes and adjustments to the recommended protocols to fit your specific needs, budget, and expertise.

The focus of this book, and the various examples that are provided, deals with studies of environmental effects from atmospheric deposition of acidifying, eutrophying, and toxic air contaminants. However, the principles that are developed and illustrated here also apply to the study of other water quality issues besides atmospheric deposition. The protocols discussed in this book can also inform the study of agricultural, silvicultural, and urban pollutants and other aspects of nonpoint source (and in some cases also point source) water pollution. A reader who grasps the materials presented in this book will be well equipped to design, implement, and interpret most types of water quality study.

We hope that the information presented here will help you to design, conduct, and interpret water quality studies that will help all of us better understand the impacts of human activities on watershed health. Armed with high-quality data and their analyses, we will collectively be able to move forward to reduce unacceptable human-caused impacts and protect and improve the quality of our water and watershed resources for future generations. Happy sampling!

# Acknowledgments

Material presented in this book is based in large part on protocols that were prepared by E&S Environmental Chemistry, Inc., under contract to the US Forest Service, Air Resources Program, under the direction of A. Mebane. Her assistance is greatly appreciated. Development of this material benefited greatly from extensive review comments received on drafts of the Forest Service protocols provided by Forest Service staff. Nevertheless, the Forest Service has not been involved in preparation of this book, and no official agency endorsement is implied. We are very grateful to Greg Lawrence for technical assistance regarding all aspects of this book. We also thank J. Charles for technical assistance in the process of preparing this book. The cover photo was taken by J.R. Webb.

# About the Authors

**Dr. Timothy Sullivan** holds a BA in history from Stonehill College, Easton, Massachusetts (1972); an MA in biology from Western State College, Gunnison, Colorado (1977); and a PhD in biological sciences from Oregon State University, Corvallis (1983) through an interdisciplinary program that included areas of focus in ecology, zoology, and environmental chemistry. He did his postdoctoral research at the Center for Industrial Research in Oslo, Norway, on surface and groundwater acidification, episodic hydrologic processes, and aluminum biogeochemistry. His expertise includes the effects of air pollution on aquatic and terrestrial resources, watershed analysis, critical loads, ecosystem services, nutrient cycling, aquatic acid-base chemistry, episodic processes controlling surface water chemistry, and environmental assessment. He has been president of E&S Environmental Chemistry, Inc., since 1988 and E&S Environmental Restoration, Inc., since 1996. He has served as project manager or lead author for a wide variety of projects that have synthesized for diverse audiences complex air and water pollution effects science. He was project manager of the effort to draft a scientific summary and Integrated Scientific Assessment (ISA) of the effects of nitrogen and sulfur oxides on terrestrial, transitional, and aquatic ecosystems for the US Environmental Protection Agency (EPA) in support of its review of the National Ambient Air Quality Standards (NAAQS). He was author of the National Acid Precipitation Assessment Program (NAPAP) State of Science and Technology Report on past changes in surface water acid-base chemistry throughout the United States from acidic deposition. He served as project manager for preparation of air quality reviews for national parks throughout California and coauthored similar reviews for the Pacific Northwest and the Rocky Mountain and Great Plains regions. He has summarized air pollution effects at all 272 Inventory and Monitoring national parks in the United States and has managed dozens of air and water pollution modeling and assessment studies throughout the United States for the National Park Service, US Forest Service, and EPA. He has published a book on the aquatic effects of acidic deposition and more than 125 peer-reviewed journal articles, book chapters, and technical reports describing the results of his research.

**Dr. Alan Herlihy** is a senior research professor at Oregon State University in the Department of Fisheries and Wildlife. His current research projects focus on developing survey design methodology, assessment approaches, and ecological indicators for assessing surface water ecological condition at large regional scales. He was a primary technical contributor to the 1990 National Acid Precipitation Assessment Program's Integrated Assessment report to Congress, the EPA/Environmental Monitoring and Assessment Program (EMAP) Mid-Atlantic Highlands Stream Assessment, and the Western Streams and Rivers Assessment. He was the primary author of the chapters in the EMAP stream and river field manual on water chemistry sampling, qualitative site assessment, and sample reach layout. Currently, he is involved with the data analysis and assessment of the National Aquatic Resource Surveys conducted by the EPA Office of Water.

**James Webb** holds a BS in environmental science from Davis and Elkins College, Elkins, West Virginia (1983) and an MS in environmental science from the University of Virginia, Charlottesville (1988). He is presently a senior scientist in the Department of Environmental Sciences at the University of Virginia, where he served for 25 years as projects coordinator for the Shenandoah Watershed Study and the Virginia Trout Stream Sensitivity Study (http://swas.evsc.virginia.edu). He has authored and coauthored numerous reports and journal articles concerning watershed response to atmospheric acidic deposition in the forested mountains of the central Appalachian Mountains region. He has participated in the design and management of water quality studies and surveys, monitoring programs, and assessments related to national park, national forest, and other conservation lands in the region. He served as an academic community representative on the Technical Oversight Committee and as coauthor of the *Aquatic Effects Technical Report* for the multiagency Southern Appalachian Mountains Initiative. He has contributed to multiple National Park Service assessments, including participation as coprincipal investigator for the Shenandoah National Park: Fish in Sensitive Habitats project and the Assessment of Air Quality and Air Pollutant Impacts in the Shenandoah National Park. He was a coauthor of the Trout Unlimited report, *Current and Projected Status of Coldwater Fish Communities in the Southeastern US in the Context of Continued Acid Deposition*. He has contributed to EPA reports on status and changes in the acid-base chemistry of surface waters in the United States related to implementation of the Clean Air Act. His previous involvement in protocol and standard operating procedure development for environmental monitoring has included work for the National Park Service and the US Department of Agriculture Forest Service.

# Acronyms and Abbreviations

| | |
|---|---|
| Al | aluminum |
| $Al^{3+}$ | free Al (uncomplexed, trivalent) |
| $Al_i$ | inorganic monomeric aluminum |
| $Al_m$ | total monomeric Al |
| $Al_o$ | nonlabile monomeric (presumed organically complexed) Al |
| $Al(F)_2^+$ | an aluminum fluoride molecule |
| $AlF^{2+}$ | an aluminum fluoride molecule |
| $Al(OH)^{2+}$ | an aluminum hydroxide molecule |
| $Al(OH)_2^+$ | an aluminum hydroxide molecule |
| ANC | acid-neutralizing capacity |
| AQRV | air quality-related value |
| ARM | Air Resource Management program |
| ARML | Air Resource Management Laboratory |
| BC | base cation |
| BMP | best management practices |
| BS | base saturation of the soil |
| $Ca^{2+}$ | calcium |
| CEC | cation exchange capacity |
| CL | confidence level |
| $Cl^-$ | chloride |
| CPR | cardiopulmonary resuscitation |
| CQCCS | calculated quality control check sample |
| DIC | dissolved inorganic carbon |
| DIW | deionized water |
| DO | dissolved oxygen |
| DOC | dissolved organic carbon |
| DON | dissolved organic nitrogen |
| DQO | data quality objective |
| ELS | Eastern Lakes Survey |
| EMAP | Environmental Monitoring and Assessment Program |
| EPA | US Environmental Protection Agency |
| F | fluorine |
| FS | US Forest Service |
| GIS | geographical information system |
| GPS | global positioning system |
| $H^+$ | hydrogen ion |
| HCl | hydrochloric acid |

| | |
|---|---|
| $HCO_3^-$ | bicarbonate |
| HDPE | high-density polyethylene |
| Hg | mercury |
| JHA | job hazard analysis |
| $K^+$ | potassium |
| LDPE | low-density polyethylene |
| LIMS | Laboratory Information Management System |
| LTM | long-term monitoring |
| μeq/L | microequivalents per liter |
| μg/L | micrograms per liter |
| μM | micromoles per liter |
| μS/cm | microsiemens per centimeter |
| m | meter |
| MAGIC | Model of Acidification of Groundwater in Catchments; a watershed ion balance model |
| $Mg^{2+}$ | magnesium |
| ml | milliliters |
| MPV | most probable value |
| N | nitrogen |
| $Na^+$ | sodium |
| NAD | North American Datum |
| $NH_4^+$ | ammonium |
| NLA | National Lake Assessment |
| $NO_3^-$ | nitrate |
| NRIS | National Resource Information System |
| NSWS | National Surface Water Survey |
| NTU | nephelometric turbidity units |
| P | phosphorus |
| PCB | polychlorinated biphenyl |
| PCU | platinum cobalt units |
| PFD | personal flotation device |
| PnET-BGC | Photosynthesis and Evapotranspiration–Biogeochemistry model; a model of water, carbon, and nitrogen balance coupled with a biogeochemistry model |
| ppm | parts per million |
| QA/QC | quality assurance/quality control |
| QCCS | quality control check sample |
| S | sulfur |
| SBC | sum of the base cations |
| Si | silicon |
| $SO_4^{2-}$ | sulfate |
| SOP | standard operating procedure |
| SRP | soluble reactive phosphorus |
| SSN | sample serial number |
| STAR | Science to Achieve Results |

| | |
|---|---|
| STORET | STOrage RETrieval (water quality database maintained by the US EPA) |
| TIME | Temporally Integrated Monitoring of Ecosystems program |
| TOC | total organic carbon |
| TSS | total suspended solids |
| USFS | US Forest Service |
| WSS | Wadeable Stream Survey |
| WTRS | surface water sample depth zone |
| lake/stream | water surface and water subsurface |

# Glossary

**acid anion:** Negatively charged ion that does not react with hydrogen ion in the pH range of most natural waters.

**acid-base chemistry:** The reaction of acids (proton donors) with bases (proton acceptors). In the context of this report, this means the reactions of natural and anthropogenic acids and bases, the result of which is described in terms of pH and acid-neutralizing capacity of the system.

**acid-neutralizing capacity (ANC):** The equivalent capacity of a solution to neutralize strong acids. The components of ANC include weak bases (carbonate species, dissociated organic acids, alumino-hydroxides, borates, and silicates) and strong bases (primarily $OH^-$). ANC can be measured in the laboratory by the Gran titration procedure or defined as the difference in the equivalent concentrations of the base cations and the mineral acid anions. It is a key indicator of the ability of water to neutralize the acid or acidifying inputs it receives. This ability depends largely on associated biogeophysical characteristics.

**acidic deposition:** Transfer of acids and acidifying compounds from the atmosphere to terrestrial and aquatic environments via rain, snow, sleet, hail, cloud droplets, particles, and gas exchange.

**acidic lake or stream:** A lake or stream in which the acid-neutralizing capacity is less than or equal to zero.

**acidification:** The decrease of acid-neutralizing capacity in water or base saturation in soil caused by natural or anthropogenic processes.

**acidified:** Pertaining to a natural water that has experienced a decrease in acid-neutralizing capacity or a soil that has experienced a reduction in base saturation.

**algae:** Photosynthetic, often microscopic and planktonic, organisms occurring in marine and freshwater ecosystems.

**algal bloom:** A reproductive explosion of algae in a lake, river, or ocean.

**alpine:** The biogeographic zone made up of slopes above the tree line characterized by the presence of rosette-forming herbaceous plants and low, shrubby, slow-growing woody plants.

**analyte:** A chemical species that is measured in a water sample.

**anion:** A negatively charged ion.

**anthropogenic:** Of, relating to, derived from, or caused by humans or related to human activities or actions.

**atmosphere:** The gaseous envelope surrounding Earth. The dry atmosphere consists almost entirely of nitrogen and oxygen, together with trace gases, including carbon dioxide and ozone.

**autecology:** Study of the ecology of individual species, as opposed to the entire community of species.

**base cation:** An alkali or alkaline earth metal cation ($Ca^{2+}$, $Mg^{2+}$, $K^+$, $Na^+$).

**base saturation:** The proportion of total soil cation exchange capacity that is occupied by exchangeable base cations, that is, by $Ca^{2+}$, $Mg^{2+}$, $K^+$, and $Na^+$.

**benthic macroinvertebrates:** Animals without backbones that inhabit the bottom substrates of streams.

**bioaccumulation:** The phenomenon wherein toxic elements are progressively amassed in greater quantities as individuals farther up the food chain ingest matter containing those elements.

**biological effects:** Changes in biological (organismal, populational, community-level) structure or function in response to some causal agent; also referred to as biological response.

**calibration:** Process of checking, adjusting, or standardizing operating characteristics of instruments or coefficients in a mathematical model with empirical data of known quality. The process of evaluating the scale readings of an instrument with a known standard in terms of the physical quantity to be measured.

**catchment:** An area that collects and drains rainwater (also called a watershed).

**cation:** A positively charged ion.

**cation exchange capacity:** The total exchangeable cations that a soil can adsorb.

**chronic acidification:** The decrease of acid-neutralizing capacity in a lake or stream over a period of decades or longer, generally in response to gradual leaching of ionic constituents.

**circumneutral:** Close to neutrality with respect to pH (neutral pH = 7); in natural waters, pH 6–8.

**climate:** Climate in a narrow sense is usually defined as the "average weather" or, more rigorously, as the statistical description in terms of the mean and variability of relevant quantities over a period of time ranging from months to thousands or millions of years. These quantities are most often surface variables such as temperature, precipitation, and wind. Climate in a wider sense is the state, including a statistical description, of the climate system. The classical period of time is 30 years, as defined by the World Meteorological Organization (WMO).

**coarse stream substrate:** Cobble- to boulder- (tennis-ball-to-car size; 64 to 4000 mm) size substrate.

**critical load:** A quantitative estimate of an exposure to one or more pollutants below which significant harmful effects on specified sensitive elements of the environment do not occur according to present knowledge.

**decomposition:** The microbially mediated reaction that converts solid or dissolved organic matter into its constituents (also called decay or mineralization).

**dissolved inorganic carbon:** The sum of dissolved (measured after filtration) carbonic acid, bicarbonate, and carbonate in a water sample.

**dissolved organic carbon:** Organic (derived from the breakdown of plant or animal material) carbon that is dissolved or unfilterable (0.45-μm pore size) in a water sample.

**drainage lake:** A lake that has a permanent surface water inlet and outlet.

**ecosystem:** The interactive system formed from all living organisms and their abiotic (physical and chemical) environment within a given area. Ecosystems cover a hierarchy of spatial scales and can make up the entire globe, biomes at the continental scale, or small, well-circumscribed systems such as a small pond.

**epilimnion:** The layer of water in a thermally stratified lake that lies above the thermocline, is circulating, and remains perpetually warm.

**episodic acidification:** The short-term decrease of acid-neutralizing capacity from a lake or stream. This process has a timescale of hours to weeks and is usually associated with hydrological events.

**EPT Index:** Index of taxonomic richness of three insect orders (Ephemeroptera-Plecoptera-Tricoptera).

**eutrophication:** The process whereby a body of water becomes overenriched in nutrients, resulting in increased productivity (of algae or aquatic plants) and sometimes also decreased dissolved oxygen levels.

**evapotranspiration:** The process by which water is returned to the air through direct evaporation or transpiration by vegetation.

**fine/sand stream substrate:** Stream substrate not gritty (silt/clay/muck < 0.06-mm diameter) to gritty (up to ladybug size; 2-mm diameter) substrate.

**glide:** Water moving slowly along stream channel, with a smooth, unbroken surface; low turbulence.

**Gran analysis:** A mathematical procedure used to determine the equivalence points of a titration curve for acid-neutralizing capacity.

**gravel stream substrate:** Fine-to-coarse gravel (ladybug to tennis ball size; 2- to 64-mm diameter) size substrate.

**groundwater:** Water in a saturated zone within soil or rock.

**hindcast:** To estimate the probability of some past event or condition as a result of rational study and analysis of available data.

**hydrologic(al) event:** Pertaining to increased water flow or discharge resulting from rainfall or snowmelt.

**hydrologic flow paths:** Surface and subsurface routes by which water travels from where it is deposited by precipitation to where it drains from a watershed.

**hydrology:** The science that studies the waters of the earth—their occurrence, circulation, and distribution; their chemical and physical properties; and their reaction with their environment, including their relationship to living things.

**hypolimnion:** The layer of water in a thermally stratified lake that lies below the thermocline, is noncirculating, and remains perpetually cold.

**Index of Biotic Integrity (IBI):** Provides assessment of biological condition, based on a combination of metrics.

**index period:** A relatively narrow period of time for synoptic sampling (often a 2-month window during spring, summer, or fall) intended to represent the lake or stream chemistry for that year.

**invasive species:** A species aggressively expanding its range and population density into a region in which it is not native, often through outcompeting or otherwise dominating native species.

**labile monomeric aluminum:** Operationally defined as aluminum that does not pass through a cation exchange column; assumed to represent inorganic monomeric aluminum ($Al_i$).

**leaching:** The removal of soil elements or applied chemicals by water movement through the soil.

**littoral zone:** Shallow-water zone around the lake margin.

**macrophyte:** Rooted aquatic plant.

**MAGIC:** Model of Acidification of Groundwater in Catchments; a watershed ion balance model.

**mitigation:** Generally described as amelioration of adverse impacts caused by a stressor such as acidic deposition at the source (e.g., emissions reductions) or the receptor (e.g., lake liming).

**model:** An abstraction or representation of a system, generally on a smaller scale.

**monomeric aluminum:** Aluminum that occurs as a free ion ($Al^{3+}$); simple inorganic complexes such as $Al(OH)_n^{3-n}$, $AlF_n^{3-n}$; or simple organic complexes, but not in polymeric forms; operationally, extractable aluminum measured by the pyrocatechol violet method or the methyl-isobutyl ketone method (also referred to as the "oxine" method) is assumed to represent total monomeric aluminum. Monomeric aluminum can be divided into labile and nonlabile components using a cation exchange column.

**nonlabile monomeric aluminum:** Operationally defined as aluminum that passes through a cation exchange column and is then measured by one of the two extraction procedures used to measure monomeric aluminum; assumed to represent organic monomeric aluminum ($Al_o$).

**occult deposition:** The removal of gases and particles from the atmosphere to surfaces by fog or mist.

**organic acids:** Heterogeneous group of acids generally possessing a carboxyl (–COOH) group or phenolic (C–OH) group.

**parameter:** (1) A characteristic factor that remains at a constant value during the analysis or (2) a quantity that describes a statistical population attribute.

**pH:** The negative logarithm of the hydrogen ion activity. The pH scale is generally presented from 1 (most acidic) to 14 (most alkaline); a difference of one pH unit indicates a 10-fold change in hydrogen ion activity.

**phytoplankton:** The plant-like forms of plankton. These single-cell organisms are the principal agents of photosynthetic carbon fixation in some freshwaters.

**plankton:** Small (often microscopic) plant-like or animal species that spend part or all of their lives in open water.

**PnET-BGC:** Photosynthesis and Evapotranspiration–Biogeochemistry Model; a model of water, carbon, and nitrogen balance coupled with a biogeochemistry model.

**pool:** In ecological systems, the supply of an element or compound, such as exchangeable or weatherable cations or adsorbed sulfate, in a defined component of the ecosystem.

**population:** For the purpose of this report, (1) the total number of lakes or streams within a given geographical region or the total number of lakes or streams with a given set of defined chemical, physical, or biological characteristics; or (2) an assemblage of organisms of the same species inhabiting a given ecosystem.

**precision:** A measure of the capacity of a method to provide reproducible measurements of a particular analyte (often represented by variance).

**primary productivity:** All forms of production accomplished by plants.

**quality assurance:** A system of activities for which the purpose is to provide assurance that a product (e.g., database) meets a defined standard of quality with a stated level of confidence.

**quality control:** Steps taken during sample collection and analysis to ensure that data quality meets the minimum standards established in a quality assurance plan.

**rapid:** Water movement along a stream channel is fast and turbulent; surface with intermittent "white water" with breaking waves; continuous rushing sound.

**reachwide sample:** All kick net samples collected at the 11 transects combined into a single composite sample.

**riffle:** Water moving along a stream channel, with small ripples, waves, and eddies; waves not breaking, and surface tension is not broken; "babbling" or "gurgling" sound.

**scenario:** One possible deposition sequence following implementation of a control or mitigation strategy and the subsequent effects associated with this deposition sequence.

**sensitivity:** For this report, the degree to which a system is affected, either adversely or beneficially, by an effect of nitrogen oxide or sulfur oxide pollution (e.g., acidification, N-nutrient enrichment, etc.). The effect may be direct (e.g., a change in growth in response to a change in the mean, range, or variability of N deposition) or indirect (e.g., changes in growth caused by the direct effect of N consequently altering competitive dynamics between species and decreased biodiversity).

**signal-to-noise ratio:** The ratio of the variance in the signal that you wish to detect (site-to-site differences or temporal changes) to the variance in the noise (variability from laboratory measurements, field crew differences, or temporal changes within the sampling window).

**species richness:** The number of species occurring in a given ecosystem, generally estimated by the number of species caught and identified using a standard sampling regime.

**specific conductance:** The conductivity between two plates with an area of $1 \text{ cm}^2$ across a distance of 1 cm at 25°C. Provides an index of the ionic strength of a water sample.

**steady state:** The condition that occurs when the sources and sinks of a property (e.g., mass, volume, concentration) of a system are in balance (e.g., inputs equal outputs; production equals consumption).

**stream flow:** Water flow within a river channel, for example, expressed in cubic meters per second ($m^3/s$) or cubic feet per second (cfs). A synonym for river discharge.

**subpopulation:** Any defined subset of the target population.

**support reach:** The length of stream to be sampled at a sampling location.

**synoptic survey:** A survey to represent water quality across a group of lakes or streams within a particular region.

**total monomeric aluminum:** Operationally defined simple unpolymerized form of aluminum present in inorganic or organic complexes.

**turnover:** The interval of time in which the density stratification of a lake is disrupted by seasonal temperature variation, resulting in an entire water mass becoming mixed.

**variable:** A quantity that may assume a numeric value during analysis.

**watershed:** The geographic area from which surface water drains into a particular lake or point along a stream.

**X-site:** Stream-sampling location.

**zooplankton:** The animal forms of plankton. Zooplankton include crustaceans, rotifers, pelagic (open-water) insect larvae, and aquatic mites.

# Chapter 1

## Purpose and Study Design

### 1.1 BACKGROUND

Scientific researchers, resource managers, and water quality practitioners often become involved in the design, implementation, and interpretation of water quality studies. Such studies take many forms, including characterization of water quality conditions, monitoring to ascertain changes in condition over time, modeling of past or future changes in water quality in response to changes in key natural and human-caused drivers of water condition, and estimation of the amounts of pollutants that can trigger biological harm. These studies focus on a host of water pollutants from a variety of pollution source types. There are point sources of sulfur (S), nitrogen (N), and mercury (Hg) from power plants and industrial facilities and diffuse nonpoint sources of N and pesticides from agricultural development. There are emissions of N from motor vehicles. Some pollutants are introduced from the land, such as nutrients from an effluent pipe flowing from a wastewater treatment facility. Other pollutants affect lakes and streams via atmospheric transport of pollutants from the sources to the location of downwind sensitive soil, vegetative, and aquatic receptors. This book focuses on the study of pollutants derived from atmospheric transport and subsequent deposition from the atmosphere to the ground surface. These pollutants change the chemistry of soils and can affect the health of plant foliage and roots. Some also move through the soil and change the chemistry of drainage water, with consequent impacts on fish and other aquatic life-forms.

Atmospherically deposited substances can acidify soil and drainage waters, alter the nutrient balance of soils or the chemistry of lakes and streams (collectively referred to as surface waters), or cause direct toxicity to aquatic biota. Some contaminants bioaccumulate in food chains, reaching high and potentially damaging concentrations in top predators, including some fish and the wildlife that feed on them.

---

### ACIDIFICATION

The potentially acidifying air pollutants that are considered here are primarily S and N. Each of these pollutants is emitted into the atmosphere in the process of burning fossil fuels, especially coal for energy production. Other potentially important anthropogenic emission sources of S and N include motor vehicles, agriculture, and other industrial sources. There are also some natural sources of S and N. These stressors contribute to multiple kinds of ecosystem effects. Both atmospheric S and N have the potential to cause acidification. Atmospheric N can also cause eutrophication of aquatic ecosystems in which the N supply limits the growth of algae or macrophytes.

Large areas throughout the United States contain substantial populations of lakes and streams having low acid-neutralizing capacity (ANC) and are therefore potentially sensitive to acidification from atmospheric deposition of S or N. These include much of the Northeast, Appalachian Mountains, northern Florida, Upper Midwest, and mountainous portions of the western United States. The eastern states include many acidified surface waters that have been affected by acidic deposition. The western states contain many surface waters susceptible to potential acidification effects, but the levels of acidic deposition in the West are relatively low, and acidic surface waters are rare. Many of the areas having acid-sensitive surface waters, especially in the northeastern United States and Appalachian Mountains, also contain extensive areas with acid-sensitive soils.

---

The study of water quality conditions and the human-caused pollutants that degrade water quality is a complex endeavor. One must consider how to collect samples that represent the conditions of a given water body of interest, analyze the samples in a laboratory, quality assure the data, analyze and interpret the resulting database, and characterize affected aquatic biological assemblages. This book provides guidelines telling you how to do that. We offer both general and highly specific recommendations regarding study design, implementation, and interpretation. More important, we explain the thinking behind the recommendations to equip you, the reader, to make your own decisions regarding how to proceed.

The focus of this book is on human-caused pollutants that acidify soils and drainage waters, those that alter nutrient availability and cycling, and toxic substances. The last include those that biomagnify in food chains. Particular emphasis is placed on effects from atmospheric inputs of S, N, and toxic metals and pesticides. Nevertheless, the principles outlined here also apply to other aspects of water quality study, including those focused on atmospherically deposited contaminants and on land-based pollution sources.

The first step in designing a surface-water-sampling program is to identify one or more problems or questions that require information on water quality. Common water quality problems include nutrient enrichment (from a variety of causes); effects of atmospheric deposition (acidification, eutrophication, toxicity); and effects of major disturbances, such as fire or pest infestations. Once the problems or questions have been clearly defined, a sampling program can be designed that addresses what to measure and where and when and how to conduct the sampling. The selection of measurements should be tailored to specific study objectives and to the study design, which guides the specifics of field, laboratory, and data analysis approaches.

## NUTRIENT ENRICHMENT

In N-limited and N- and P-colimited aquatic systems, atmospheric N deposition can influence algal growth, trophic state, and the distribution and abundance of diatoms and perhaps other aquatic species. Lakes and streams that are wholly or partly N limited are most likely to occur in remote regions with naturally oligotrophic surface waters that have not received high levels of atmospheric N deposition in the past. Within the United States, such lakes and streams are most common in the mountainous West.

A variety of air pollutants have the potential to stress aquatic ecosystems through contributions from the atmosphere to Earth's surface. A major focus of this book is on atmospheric pollutants that contribute to surface water acidification and eutrophication (nutrient enrichment). Both atmospheric S and N have the potential to cause acidification. Atmospheric N can also cause eutrophication of aquatic ecosystems in which the N supply is limiting for algal or plant growth. Sampling for atmospherically deposited toxic materials is also addressed, but in less detail.

Many freshwaters in the United States are thought to be phosphorus (P) limited (US Environmental Protection Agency [EPA] 2008). In such waters, addition of P would be expected to increase plant or algal growth, whereas addition of N would not. Nevertheless, there are also freshwaters considered to be N limited or N and P colimited. Furthermore, nutrient limitation can vary with season.

Individual researchers and government entities collect and analyze data on natural resource sensitivity to, and effects from, air pollution on aquatic and terrestrial ecosystems nationwide. Such studies are often conducted at the individual park, forest, or wilderness level. Larger regional and even national studies are also conducted. This book provides a consistent framework regarding decisions of where, when, and how to conduct water sampling for the purpose of evaluating and monitoring air pollution effects on aquatic ecosystems. It also describes how to conduct laboratory analyses, how to quality assure project data, and how to analyze and interpret the database developed in a water quality study. It is based on protocols developed for the Forest Service (FS) by Sullivan (2012), which in turn is based on protocols previously developed by or for the US EPA, US Geological Survey (USGS), and FS, including documents prepared by Herlihy (1997), Turk (2001), Webb et al. (2004), Eilers (2007), and Sullivan and Herlihy (2007). Appendix A provides a list of protocols, guidance documents, and methods manuals that were reviewed.

This framework allows the user to build a site-specific project plan based on relevant research and management questions. References that provide more details on these measurements are also included. Individual chapters address approaches for water sampling, laboratory analyses, quality assurance/quality control (QA/QC), data analysis, and approaches for sampling and analyzing aquatic biota. A final chapter addresses issues that arise when you transition from one set of methods to another.

## 1.1.1   Resources Sensitive to Atmospheric Deposition

The most common ecological air quality-related values (AQRVs) that are susceptible to air quality degradation are water (and associated aquatic fauna), soil, and flora. Sensitive receptors for effects on surface water include water chemistry, productivity, and the response of life-forms, including fish, zooplankton, benthic macroinvertebrates, and phytoplankton. This book focuses mainly on surface water chemistry, and secondarily on zooplankton and benthic macroinvertebrates.

AQRVs are resource elements that could be damaged by air pollution or atmospheric deposition. There are many possible sensitive indicators for each AQRV. To protect the AQRV water, sensitive receptors might include the chemistry of the water, which could influence its suitability to support various aquatic species and life-forms. ANC is an indicator of change for the sensitive receptor water chemistry. There are also sensitive biological receptors, which reflect the suitability of the lake water for supporting aquatic organisms that might be sensitive to acidification or eutrophication. These could include, for example, specific species of fish, zooplankton, insects, or diatoms. A sensitive receptor can be evaluated by measuring indicators of injury or ecosystem change.

There are many approaches that can be used to assess (1) current condition of surface waters; (2) the sensitivity of aquatic natural resources to potential degradation from atmospheric deposition of S, N, or toxic materials; and (3) the extent to which sensitive aquatic natural resources have been harmed in the past or might be expected to be harmed in the future under assumed scenarios of future air pollution and atmospheric deposition. Site-specific studies can be further customized to fit particular regional or local ecosystem conditions and stressors.

Atmospheric deposition can contribute to toxicity responses in several ways. Water acidification entails several chemical changes. These include reduced pH (increased hydrogen ion [$H^+$] activity), decreased ANC, increased inorganic monomeric aluminum ($Al_i$) concentration, and changed (increased or decreased, depending on the extent of acidification) concentrations of calcium ($Ca^{2+}$) and other base cations (BCs). Hydrogen ion and $Al_i$ can be toxic to many aquatic species at sufficiently high concentrations. Other atmospheric pollutants of concern with respect to toxicity include Hg and various pesticides. Atmospheric deposition is an important component of Hg cycling and biogeochemistry. Mercury in its methylated form (MeHg) is known to bioaccumulate in aquatic organisms, reaching potentially high concentrations in larger, piscivorous fish and species that consume them.

A limited list of key variables does not exist with which to measure ecosystem condition, or ecosystem response to stressors, such as those associated with atmospheric deposition (i.e., acidification, eutrophication, toxicity). Ecosystems are highly complex and simply cannot be represented by a handful of variables. Nevertheless, there are variables that have been shown to be, or that are expected to be (based on existing research), reflective of the general level of ecosystem harm that might be associated with atmospheric deposition (Table 1.1; Sullivan and Herlihy 2007). The recommended AQRVs and sensitive receptors summarized here are broadly applicable and reflect a range of aquatic effects of atmospheric deposition. Identification of these receptors and indicators helps determine the approaches that will be needed for inventory and monitoring.

Detailed protocols should be an important part of any resource characterization or monitoring program intended to evaluate atmospheric deposition impacts on AQRVs. Standardized approaches help to ensure that measured differences among locations or changes over time at one location are real (actually occur in nature) and are not simply a reflection of different methods, sampling personnel, or timing of sample collection. Protocols are necessary to ensure that the data collected are appropriate to the questions asked and are of sufficient quality to allow development of meaningful answers. It must, however, be recognized that there will not be a single appropriate approach in every situation that will efficiently characterize an important attribute nationwide. Some attributes and site characteristics are sufficiently variable from region to region so that supplemental or amended protocols may be justified. Nevertheless, adoption of standardized procedures for data gathering and analysis and required core data elements will allow data to be compared across regional boundaries and will

**TABLE 1.1    RECOMMENDED AQRVs, SENSITIVE RECEPTORS, AND INDICATORS AFFECTED BY ATMOSPHERIC DEPOSITION OF AIR POLLUTANTS**

| AQRV | Sensitive Receptor | Indicator/Metric | Potential Criteria[a] |
|---|---|---|---|
| Flora | Red spruce (East) | Growth<br>Decline | Change in diameter<br>Change in extent of damage |
| | Sugar maple (East) | Growth<br>Decline | Change in diameter<br>Change in extent of damage |
| | Lichens | Community composition | Loss of sensitive taxa |
| Soil | Soil chemistry | Base saturation (BS)<br>Exchangeable $Ca^{2+}$<br>Exchangeable $Ca^{2+}$ + $Mg^{2+}$<br>C:N molar ratio | BS < 10%<br>% change over time<br>% change over time<br>C:N < 0.2 |
| | Soil solution chemistry | Ca:Al molar ratio<br>$[Ca^{2+} + Mg^{2+} + K^+]$:Al molar ratio<br>$NO_3^-$ concentration | Ca:Al < 1.0<br>BC:Al < 1.0<br><br>$NO_3^-$ > 20 µeq/L during growing season |
| Water | Water chemistry | ANC<br>$NO_3^-$ concentration<br>$SO_4^{2-}$ concentration | ANC < 50 µeq/L<br>$NO_3$ > 10 µeq/L<br>Change over time |
| | Water productivity | Chlorophyll *a*<br>Clarity (lakes) | Change over time<br>Change over time |
| | Fish | Salmonid species presence<br>Fish species richness<br>Fish condition factor<br>Fish Hg concentration<br>Fish pesticides concentration | Loss over time<br><br>Change over time<br>Change over time<br>Hg > 0.3 ppm<br>Above threshold values |
| | Zooplankton (lakes) | Total zooplankton richness<br>Crustacean taxonomic richness<br>Rotifer taxonomic richness | Change over time<br><br>Change over time<br>Change over time |

**TABLE 1.1 (*Continued*)   RECOMMENDED AQRVs, SENSITIVE RECEPTORS, AND INDICATORS AFFECTED BY ATMOSPHERIC DEPOSITION OF AIR POLLUTANTS**

| AQRV | Sensitive Receptor | Indicator/Metric | Potential Criteria[a] |
|---|---|---|---|
| | Benthic macroinvertebrates (streams) | Mayfly taxonomic richness<br>Index of Biotic Integrity | Loss of sensitive taxa<br>Deviation from reference |
| | Diatoms | Community composition | Historical change from paleolimnological reconstruction |

*Source:* Modified from Sullivan, T.J., and A.T. Herlihy. 2007. *Air Quality Related Values and Development of Associated Protocols for Evaluation of the Effects of Atmospheric Deposition on Aquatic and Terrestrial Resources on Forest Service Lands.* Final report prepared for the USDA Forest Service. E&S Environmental Chemistry, Corvallis, OR.

[a] Metrics can be represented in multiple ways, often as change over time detected in a monitoring program or as exceedance above or below a threshold value. Typically, multiple threshold values are possible. For example, surface water target ANC thresholds are commonly set at 0, 20, or 50 µeq/L to achieve different levels of protection.

provide information that is needed for national assessments and decision making. Aquatic effects inventory and monitoring for atmospheric deposition effects in the United States have historically focused on both lakes and streams. To the extent practical, this book describes attributes and methods that are applicable to both lakes and streams and that can be applied throughout most or all regions of the United States. This book addresses which of the sensitive surface water receptors, and associated field, laboratory, and data analysis approaches, are most useful for meeting specified objectives.

### 1.1.1.1 Sensitive Chemical Indicators of Water Quality

Acidification sensitivity and level of effect are commonly evaluated using several chemical criteria, especially ANC, pH, and $Al_i$. Sensitivity of surface waters to chronic and episodic acidification depends on watershed soil characteristics, mineralogy, and hydrologic flow paths within the watershed (Chen et al. 1984, Cosby et al. 1985), as well as on the current and historic atmospheric deposition loadings of acids and bases. Surface water ANC provides an initial point of departure from which to assess quantitatively the current status of stream or lake chemistry. Biological effects of acidification have been associated with a variety of ANC benchmarks, the most common of which are ANC equal to 0, 20, and 50 microequivalents per liter (µeq/L). ANC of 0 µeq/L or less is of significance because waters at or below this level have no capacity to neutralize acid inputs; they are acidic by definition. Lakes and streams with ANC chronically below 0 µeq/L are often fishless or contain few species of fish. The relatively acid-tolerant species, brook trout (*Salvelinus fontinalis*), has been shown to be sensitive to episodic acidification[*] when chronic stream ANC is below about 20 µeq/L. A general benchmark for sensitivity of other types of aquatic biota is often established at ANC equal to 50 µeq/L (Driscoll et al. 2001). Some species may be affected at higher ANC values, even at levels of 100 µeq/L or above. However, model hindcast simulations suggest that many lakes and streams did not have ANC that high prior to the Industrial Revolution and initiation of acidic deposition. Generally, surface waters with ANC of 50 µeq/L or less are considered prone to episodic acidification in some regions (DeWalle et al. 1987, Eshleman 1988), especially where seasonal snowpack accumulations are substantial. Such low-ANC waters may also be susceptible to future chronic acidification at current or increased rates of acidic deposition.

Common reference values for pH are 5.0, 5.5, and 6.0. Such values are often used to evaluate the possible extent of adverse effects on fish and other aquatic organisms. Threshold pH levels for adverse biological effects have been summarized for a variety of aquatic organisms (Haines and Baker 1986, Baker et al. 1990). The effects of low pH are specific to the organism and region under consideration and depend on the concentrations of other chemical constituents in the water, notably $Al_i$ and $Ca^{2+}$. Lakes or streams having pH below about 5.0 generally also have ANC below 0 and often do not support fish. Depending on the region, waters having pH above about 6.5 and ANC above about 50 to 100 µeq/L support large, but variable, numbers of species. Populations of salmonid fish are generally not found at pH levels less than 5.0, and smallmouth bass (*Micropterus dolomieu*) populations are usually not found at pH values less than 5.2 to 5.5 (Haines and Baker 1986). A number of synoptic surveys[†] indicated loss of species diversity and absence of many other fish species in the pH range 5.0 to 5.5 (Haines and Baker 1986). Levels of pH less than 6.0 to 6.5 have been associated with adverse effects on populations of dace, minnows, and shiners (family Cyprinidae), and bioassays suggest that given sufficient $Al_i$ concentrations, pH less than 6.5 can lead to increased egg and larval mortality in blueback herring (*Alosa aestivalis*) and striped bass (*Morone saxatilis*; Hall 1987, Klauda et al. 1987).

---

[*] Episodic acidification refers to the temporary (typically hours to days) decrease in lake, or especially stream, ANC that occurs in response to hydrologic events such as rainfall or snowmelt.
[†] These are surveys to represent water quality across a group of lakes or streams within a particular region.

Aluminum toxicity to aquatic organisms is caused primarily by inorganic forms of Al rather than organically complexed Al ($Al_o$; Driscoll et al. 1980, Baker and Schofield 1982, Havas 1985). There is limited evidence of biological effects at $Al_i$ less than 50 µg/L (~2 µM). Free Al concentrations ($Al^{3+}$; roughly equivalent to $Al_i$ concentrations at pH values substantially below 5.0) between 50 and 200 µg/L have been shown to reduce the growth and survival of various species of fish (Muniz and Levivestad 1980, Baker and Schofield 1982). Concentrations of $Al_i$ greater than 200 µg/L are generally considered to have toxic effects on the majority of freshwater fish species (Table 1.2).

Sensitivity of surface waters to eutrophication and the nutrient status of lakes and streams are typically evaluated on the basis of concentrations of P and N. These nutrients can be assessed as total N or total P or as one or more of the various forms that commonly occur in surface waters, such as nitrate ($NO_3^-$) and soluble reactive P (SRP). The US EPA has provided guidance to states for setting nutrient criteria for total N and P concentrations in US lakes (US EPA 2000a) and streams and rivers (US EPA 2000b). Different nutrient criteria are being developed for each of 14 different nutrient ecoregions throughout the country. Nutrient ecoregions are based on aggregations of level III Omernik ecoregions (Omernik 1987; http://www.nps.gov/plants/sos/pdf/SOS%20Omernik%20Level%20III.pdf).

In some areas, the concentrations of potentially toxic substances in surface waters may be of concern. This issue is likely to be of greatest interest in areas downwind of substantial emissions sources of pesticides or where atmospheric deposition of Hg (or another trace metal) is known to be elevated. Monitoring of pesticides in surface waters may be advisable on lands directly downwind of intensive agricultural development.

Studies of Hg concentrations in fish tissue may be warranted in areas that are downwind of known Hg emissions sources, especially where such areas

**TABLE 1.2 GENERAL GUIDELINES[a] FOR EVALUATING THE LIKELIHOOD OF AL TOXICITY IN FRESHWATERS**

| Concentration of Inorganic Monomeric Al | Expected Response of Aquatic Biota |
|---|---|
| <50 µg/L | Biological effects not likely to most forms of aquatic biota |
| 50 to 200 µg/L | Reduced growth and survival of various species of fish, including brook trout, and likely other aquatic life-forms |
| > 200 µg/L | Adverse effects likely for most freshwater fish species |

[a] These are general guidelines. Variability is high with species, life stage, and various aspects of water chemistry, including $Ca^{2+}$ concentration, dissolved organic carbon (DOC), and total fluoride (F) concentration.

co-occur with probable geologic sources of Hg or areas that promote Hg methylation. Regional Hg deposition attributable to long-range atmospheric transport is also of concern. In general, we do not recommend routine monitoring for Hg concentrations in water. Nevertheless, Hg can pose a health risk to humans or wildlife (i.e., bald eagle, osprey, loon, and river otter) that consume large quantities of such fish. We recommend that a more effective way to evaluate Hg contamination issues in surface waters is to analyze, or monitor, concentrations of Hg in fish rather than in water. Of particular concern are the larger, older, piscivorous fish, such as bass, pike, and some species of trout.

### 1.1.1.2    Potential Confounding Factors

In developing and implementing a field-sampling program, it is important to consider numerous potentially confounding factors. Some of the important considerations that can complicate aquatic inventory and monitoring assessments include:

- low signal-to-noise ratio,[*] especially for dilute aquatic ecosystems,
- variation in watershed properties, such as slope, aspect, underlying bedrock composition, glacial till (if it occurs), depth and composition of soils, distribution of vegetative cover, role of groundwater, and presence and hydrologic connectedness of wetlands,
- interacting stressors, especially climate, introduced species, and legacy effects of past land use and exposure to pollutants,
- constraints of sampling in designated wilderness areas where land use rules prohibit access via mechanized equipment and installation of fixed equipment,
- constraints regarding laboratory analytical holding times,
- national and regional applicability,
- cost and training constraints, and
- quality control issues and the need for peer review.

### 1.1.2    Study Purpose and Objectives

Prior to selection of study sites and parameters to measure, it is important to determine the purpose of the sampling program. For example, the main purpose could be to

- evaluate nutrient limitation,
- document temporal variability (i.e., diurnal, episodic, seasonal, annual, interannual) in water chemistry,

---

[*] In other words, natural and sampling variability are high relative to the magnitude of change that has occurred in response to atmospheric deposition.

- evaluate the spatial extent of acid-base status,
- parameterize interpretive and predictive models,[*]
- determine sensitivity of resources to potential damage, or
- estimate magnitude of impact on water quality.

Variation in purpose dictates variation in general approach (Table 1.3), which in turn influences the selection of appropriate protocols.

The management needs that the field study is intended to address will help determine the type of field study that might be most appropriate. The management needs will lead to a series of questions, which in turn will guide the sampling effort. Such questions might include the following:

- What kinds of sampling are required to support management needs?
- What are the approaches most appropriate for meeting those sampling requirements?
- What are the standard operating procedures (SOPs) to implement those protocols?

In designing a field study, there are 10 questions that have been identified by the US EPA that should be considered (Table 1.4). Unless each of these questions is addressed, there is risk that the sampling program will fail to yield the data required to meet the program's needs. This book provides guidance regarding how to answer such questions.

The most important aspect of any inventory and monitoring plan is specification of the objectives and questions to be answered using the resulting data. Once the objectives and questions are conceived and refined and some preliminary data are collected with which to evaluate data variability issues, it is possible to specify a plan that will have a high probability of success. The greatest challenge in developing a monitoring or synoptic survey plan is asking the best questions. It is important to decide what it is that you want to know and what uncertainty you are willing to accept in your answers. Many field-sampling programs are destined to fail from the outset because they were not specific about what questions the program was intended to answer. Specificity regarding the questions can lead to specificity regarding the sampling design and result in the collection of data capable of providing answers that will help in resource management.

Because it is not possible to sample at all locations at all times for all parameters, it is important to consider in advance how to make the best choices regarding expenditure of limited funds for field sampling. The most important aspect of sampling design is setting specific objectives and linking these

---

[*] To apply a process-based effects model to a particular site, various input data are needed, depending on the selected model. Such data might include, for example, characterization of soils, hydrology, vegetation, or historical documentation of land use or atmospheric deposition.

**TABLE 1.3   COMMON MANAGEMENT ISSUES WITH ASSOCIATED FIELD STUDY APPROACHES**

| Purpose | General Approach |
|---|---|
| Determine whether one lake or stream, or a group of lakes or streams, is N limited for algal growth | Sample water and determine nutrient and chlorophyll *a* concentrations on multiple occasions (at least monthly during the snow-free season) during multiple years. Consider also nutrient (N, P) addition experiments in the laboratory or field enclosures. |
| Quantify episodic excursions from base flow conditions in surface water chemistry (i.e., ANC, pH, $Al_i$, $NO_3^-$ concentrations) during hydrologic events | Sample water and measure full ion chemistry during rainstorms, snowmelt, or rain-on-snow events, with hourly to weekly periodicity. |
| Determine the distribution of lake or stream water chemistry (i.e., ANC, pH, $NO_3^-$ concentration) across a particular study area | Conduct a statistically based or systematic synoptic survey of lake or stream chemistry. |
| Quantify long-term changes in lake or stream ANC (or other variable) over time in a particular lake or stream | Sample at least annually (preferably monthly or seasonally during the open water season) over a period of at least 8 years. Consider restricting sampling times to common hydroperiod or other approach to standardize timing of sample collection among years. Length of time required to continue monitoring to document statistically significant changes will depend on temporal variability in water chemistry and extent of long-term changes that occur. In general, at least 8 years of data will be required. |
| Determine to what extent air pollution is currently affecting the water resources in a particular study area | Multiple approaches can contribute to this evaluation, as follows: 1. Characterize index chemistry for multiple lakes or streams expected to be highly sensitive 2. Conduct synoptic survey (preferably using a stratified random selection process) of waters in the study area |

TABLE 1.3 (*Continued*)  COMMON MANAGEMENT ISSUES WITH ASSOCIATED FIELD STUDY APPROACHES

| Purpose | General Approach |
|---|---|
| | 3. Use a dynamic, process-based watershed model to hindcast past changes in acid-base chemistry |
| | 4. Collect and analyze diatom remains in a sediment core from the deepest part of one or more of the presumed most acid-sensitive lakes |
| | 5. Use a steady-state or dynamic-process-based watershed model to quantify the critical load of S or N deposition |
| Evaluate whether the current condition of acid or nutrient-sensitive waters warrants mitigation | Multiple approaches can contribute to this evaluation, as follows:<br>1. Characterize index chemistry for multiple lakes or streams expected to be highly sensitive<br>2. Conduct synoptic survey (preferably using a stratified random selection process) of waters in the study area<br>3. Use a dynamic-process-based watershed model to hindcast past changes in acid-base chemistry<br>4. Collect and analyze diatom remains in a sediment core from the deepest part of one or more of the presumed most acid-sensitive lakes<br>5. Use a steady-state or dynamic-process-based watershed model to quantify the critical load of S or N deposition<br>6. Use a dynamic-process-based model to evaluate likely future responses to reduced levels of acidic deposition |

objectives to specific questions. These questions should consider elements of subject, location, time, trend, degree, and population of interest (Table 1.5).

A well-conceived plan for water quality sampling should be (Eilers 2007)

- relevant to the intended beneficial uses of the waters,
- specific with respect to sampling locations, depths, parameters, schedule, and methods,
- consistent with approved methods,

**TABLE 1.4    TEN BASIC QUESTIONS TO CONSIDER IN DESIGNING WATER QUALITY CHARACTERIZATION, MONITORING, AND ASSESSMENT STUDIES**

1. Why is the sampling taking place?
2. Who will use the resulting data, and how will that influence the level of quality assurance that will be required?
3. How will the data be used, and how will the intended use influence data requirements?
4. What parameters or conditions will be measured?
5. How good do the data need to be in terms of accuracy, representativeness, completeness, and intrasite and intersite comparability?
6. What methods should be used?
7. Where are the sampling sites?
8. When will the sampling occur?
9. How will the data be managed?
10. How will the program ensure that the data are credible?

**TABLE 1.5    EXAMPLE OF ELEMENTS TO BE CONSIDERED IN FORMULATING SAMPLING QUESTIONS**

| Element | Example |
|---|---|
| Subject | Stream $NO_3^-$ concentration |
| Location | Spring Creek, 50 m below its confluence with Sparks Creek |
| Time | During spring snowmelt |
| Trend | Is stream $NO_3^-$ concentration increasing from year to year during the spring high-flow period? |
| Degree | Is it changing by a statistically significant amount or a biologically meaningful amount? |
| Population of Interest | First- through third-order (at 1:100,000 scale) streams in the Blue Ridge ecoregion in Virginia |

- specific with respect to recommendations for data analysis, reporting, and flagging, and
- designed to maintain continuity to the extent possible with the existing sampling efforts, especially if trend analyses will be conducted using the data.

Within the context of characterization and monitoring studies to measure or document air pollution effects on surface waters, a multitude of questions exists that could be considered appropriate for focusing field studies. These are far too numerous to list. A partial list is given in Table 1.6. Selection of the most appropriate questions depends to a large degree on location. Key questions can be influenced by the extent of historical acid and nutrient deposition; inherent sensitivity of the resources present; hydrologic characteristics; types of aquatic resources of greatest interest (i.e., drainage lakes, seepage lakes, low-order streams, moderate-order streams); topography; and others.

**TABLE 1.6    EXAMPLES OF QUESTIONS THAT COULD BE USED TO GUIDE INVENTORY, CHARACTERIZATION, AND MONITORING STUDY DESIGN**

### Inventory

What is the distribution of lake water ANC (or alternatively pH; inorganic monomeric Al, Ca, $NO_3^-$, or $SO_4^{2-}$) across high-elevation lakes in XYZ Wilderness Area?

What is the annual average (or index) water chemistry of the most acid-sensitive streams in XYZ National Park or National Forest (expressed as fifth percentile of sensitivity of the population of streams or the five most sensitive streams known to exist)?

What are the concentrations of stream water $NO_3^-$ (or ANC, pH, $Al_i$) during snowmelt at selected long-term monitoring locations, and how do they compare with summer or fall index $NO_3^-$ concentration in these streams?

### Characterization

What is the extent of episodic chemical change (decrease in ANC, pH; increase in inorganic monomeric Al, $NO_3^-$) during the peak of snowmelt at selected long-term monitoring stream sites?

What landscape characteristics (i.e., lithology, soil type, elevation, ecoregion, stream order, etc.) are associated with the occurrence of streams having spring base flow ANC below 50 μeq/L within the forests of a particular state?

### Monitoring

What is the long-term trend in lake water $NO_3^-$ (or other variable) concentration for monitoring sites in the Rocky Mountains over the period of monitoring since 1990, as measured during the summer index period, and what are the characteristics of the sites that show the largest positive trends?

Given the observed temporal variability in spring base flow ANC in a particular stream within a certain study area, how long would monitoring need to be conducted to document a statistically significant increase in stream ANC if the average actual increase in ANC was 1 μeq/L/year?

Do long-term trends in spring base flow stream water $Ca^{2+}$ concentrations in second- and third-order streams in XYZ Wilderness Area since 1990 suggest the potential for Ca deficiency in the soils of higher-elevation forests in this wilderness?

Monitoring of lake and stream water quality is performed to provide resource managers with information on possible water quality problems that may require intervention, to determine the susceptibility of lakes and streams to potential stressors, and to document future changes (improvement or deterioration) in key parameters of interest or in known problem areas. For example, resource managers may need to know surface water sensitivity to acidification when they review emissions permit applications. Information from a well-designed and properly executed monitoring plan will allow future evaluation of the effectiveness of emissions controls or other best management practices (BMPs) and the potential need for other actions that might be warranted.

## 1.2  STUDY DESIGN

Water quality studies for evaluating aquatic effects of atmospheric deposition are most commonly designed as lake or stream characterization studies, synoptic surveys of the chronic chemistry of lakes or streams in a particular forest or region, characterization of episodic variations in chemistry in response to rainstorms or snowmelt, or long-term monitoring (LTM) studies to document and quantify changes in chemistry over time. Each type of design is described in the following material. In selecting an appropriate design, it is most important to determine in advance precisely what it is that you would like to know. From this will spring logically the type of study design that will be most useful (Table 1.7).

One of the most important, and most frequently overlooked, aspects of study design is that it should incorporate the data requirements of the statistical procedures that will be used to analyze the data. Thus, it can be helpful to consult with a statistician, or at least a person who is knowledgeable about statistics, as part of the process of developing the study design. In addition, it is helpful to consider what you intend to do with the data before you design the study and before you actually collect the data. It can also be helpful to coordinate with existing efforts by other research groups or governmental agencies. This coordination effort may be as simple as collecting some additional data that might be shared or pursuing joint funding for a desired sampling effort.

### 1.2.1  Lake or Stream Characterization

To characterize the chemistry of a lake or stream, there are multiple possible approaches for water quality characterization. Some studies are based on only one or a few samples. Most commonly, these are collected as index samples. Decisions will need to be made concerning the frequency and timing of sampling. Springtime base flow samples are often regarded as a good

**TABLE 1.7    GENERAL GUIDANCE REGARDING WATER QUALITY STUDY DESIGN**

| If What You Want to Know Is | You Should Consider the Following Kind(s) of Study Design |
|---|---|
| Number of lakes, length of streams, or percentage of the regional population of lakes or stream length that is above or below a particular criteria value (i.e., ANC $\leq$ 50 $\mu$eq/L) | Some form of stratified random sampling, which will allow extrapolation of results from individual sites to the larger area |
| Status of the acid-base chemistry of the most (or some of the most) sensitive lakes or streams in an area | Nonstatistical survey of selected lakes or streams in portions of the study area and landscape positions expected to contain the most sensitive aquatic resources. |
| General assessment of lake or stream chemistry in an area, with identification of some of the more sensitive water bodies in the area | Statistical or nonstatistical screening of a relatively large number of water bodies across the expected gradient of sensitivity, measuring specific conductance or pH in the field for making a rough assessment of condition, and collecting samples for full laboratory analyses for a subset of those samples. |
| Estimate of seasonal or episodic variability in the chemistry of an acid-sensitive lake or stream | Frequent interval sampling during the period of interest (typically snowmelt or during rainstorms). Sampling can range from hourly to monthly during the season or period of interest and can include multiple years to capture the range of variation. |
| Analysis of changes in water chemistry over time | Periodic sampling (usually monthly to annual) over a period of usually at least 8 years, focused on an index period or standardized by hydroperiod. More robust studies (with greater statistical power to detect trends) will entail more frequent sampling (weekly to seasonal) or will extend for longer periods of time. |
| Assessment of temporal variability in water chemistry of a particular lake or stream | Frequent interval (hourly to seasonal) sampling that captures major changes in hydrology during the season(s) of interest. Should include multiple years. |

*Continued*

**TABLE 1.7 (*Continued*)   GENERAL GUIDANCE REGARDING WATER QUALITY STUDY DESIGN**

| If What You Want to Know Is | You Should Consider the Following Kind(s) of Study Design |
|---|---|
| Determination of whether, and to what extent, water resources in a particular area have been adversely affected by atmospheric deposition to date | Multiple designs will be needed, employing a weight-of-evidence approach. They might include the following:<br>• synoptic survey (statistical survey preferred)<br>• characterization of multiple representative sensitive lakes or streams<br>• long-term monitoring<br>• assessment of seasonal and episodic variability<br>• hindcast chemistry using dynamic-process-based model(s) |
| Determination of the prognosis for future recovery of damaged aquatic resources or quantification of the atmospheric deposition levels that will be protective of sensitive resources | Model scenario and critical loads analysis |

representation of annual average flow-weighted stream water quality when only single samples can be collected. Summer or fall index (commonly avoiding large rainstorm events) chemistry is often regarded as a good representation of annual average lake water chemistry. Lake sampling after fall overturn can yield results for fully mixed conditions but may require measuring the lake temperature profile to verify that turnover has occurred. Selection of an index period has implications for the temporal stability of the water quality and for the degree of impact that might be revealed by that water quality. Water quality is more likely to be stable (and thus comparable among water bodies if a survey is conducted of multiple lakes or streams) during summer and fall. However, in many regions, the lowest pH and ANC, and highest $Al_i$ and nutrient concentrations, are likely to occur during spring.

Better representation of annual conditions in both streams and lakes can be obtained with seasonal or other periodic sampling, as opposed to collection of only one sample to represent a given year. Selecting a sampling period when flows are low and least variable may provide data that are generally comparable among a group of waters or comparable in a given water body between years but may not represent the conditions of greatest interest. For example, surveys to assess the effects of nutrient input or acidic deposition on surface water chemistry can substantially underestimate impacts if low-flow periods are used for sampling (Lawrence et al. 2008). Some studies endeavor to collect one or more samples during high-flow periods (heavy rain or snowmelt), perhaps to augment index chemistry sampling. During high-flow periods, hydrologic conditions tend to cause

- relatively high $NO_3^-$, $Al_i$, and P concentrations,
- relatively low pH and ANC, and
- variable concentrations of $SO_4$ and BC (depending on local characteristics).

Therefore, the most stressful conditions (to aquatic biota) often occur during high-flow periods. The range of difference between high-flow and low-flow chemistry varies by region, with drought cycles, and by individual water body. In general, we recommend that characterization of lakes and streams be represented by one or more index samples for a given year, plus at least two additional samples during snowmelt or rainstorm to partially characterize variability. It is preferable to collect characterization data for at least 2 years and to document interannual variability associated with wet/drought cycles.

## 1.2.2  Synoptic Survey

Synoptic surveys of lake or stream chemistry within a designated study area are usually collected at times expected to exhibit fairly stable water chemistry. For acidic and nutrient deposition monitoring, this is usually spring base flow for streams in the southeastern United States and is the summer or fall index period[*] for lakes or streams in regions that typically develop substantial snowpack. Typically, one sample, sometimes with replicates, is collected for each lake or stream that was selected for sampling. Synoptic surveys are ideally (but not always) statistically based, allowing extrapolation of sample results for individual water bodies to the regional population of interest. It is desirable to standardize conditions at each site at the time of sample collection to generally consistent weather and runoff conditions. For that reason, periods of high temporal variability such as heavy rain and periods of rapid snowmelt, or periods with heavy smoke from wildfires, are typically avoided to the extent possible during a synoptic survey. However, if assessment of acidic or

---

[*] An index period is a relatively narrow period of time for synoptic sampling (often a 2-month window during spring, summer, or fall) intended to represent the lake or stream chemistry for that year. Typically, rain or snowmelt conditions are avoided when collecting index samples.

nutrient deposition effects is the goal, avoiding high-flow conditions can result in a substantial underestimation of the magnitude of impact. The extent of this underestimate can be quantified by conducting additional seasonal and episodic sampling for at least a subset of the sampling sites.

### 1.2.3  Characterization of Stream Chemistry during Hydrologic Events

Hydrologic events, which are high flows caused by rainstorms or rapid snowmelt, are episodic and can last from a few hours to a few weeks. Although these events occur over a relatively small fraction of the year, they often represent the majority of total annual flow (Likens et al. 1977) and constituent flux. Events can happen throughout the year in most regions of the United States but are most common during seasons of high precipitation (and during spring snowmelt in regions where snow accumulates) and are generally least common during summer, when high evapotranspiration reduces soil moisture.

Stream water chemistry during hydrological events is important to characterize because high flows often lead to extreme chemical conditions. Effects on stream biota from episodic variations in stream chemistry can be as severe or more severe than chronic effects associated with base flow chemistry. For example, episodic acidification can result in the elimination of an annual age class of fish (McComick and Leino 1999) when the event is timed with the presence of sensitive life stages. It can also affect other forms of aquatic life, such as diatom communities, which have been found to be less diverse in an episodically acidified stream than in a nearby chronically acidified stream (Passy et al. 2006).

---

### SAMPLING TO SUPPORT MODELING

If modeling is anticipated using the collected data, it is important to consider which water chemistry variables will be needed to calibrate the model, for example, application of the Model of Acidification of Groundwater in Catchments (MAGIC).

---

Manual water sampling is generally not effective at characterizing chemical variability over the course of a hydrologic event because the timing and shape of the hydrograph is difficult to predict and may occur at inconvenient times for the sampler. Automated water sampling triggered by changes in water level (often referred to as stage) provides an effective solution to this problem. Automatic samplers can collect water at preset time intervals or at intervals based on the rate of change in water level, which is usually not constant over time. Samples collected during events can then be evaluated using the flow measurements recorded at the times of sample collection to optimize the selection of samples for chemical analysis. This approach offers the opportunity to greatly reduce sampling costs with minimal loss of information. However, the use of automatic-sampling equipment is moderately expensive, and its use is restricted in wilderness settings.

### 1.2.4  Long-Term Monitoring

Long-term monitoring of stream or lake chemistry usually involves collection of water samples at regular intervals from weekly to quarterly or even annually, with the primary purpose of detecting trends that reflect an environmental change over time. How quickly a trend can be detected depends on the strength of the trend (the rate of change) and the amount of intra-annual and interannual variability in the water chemistry. It is generally possible to detect a change of smaller magnitude under conditions of less variability and longer period of record. The likelihood of detecting a significant trend in the concentration of a given water chemistry variable will be determined in large part by the length of the monitoring period. In the event of a small-to-moderate change in chemistry, it may take 10 to 20 years, or more, of monitoring data to document a significant change.

An effective monitoring plan stems from a series of questions and constraints that sequentially focus the plan into specific elements that are well defined and unambiguous. Because information is gained during implementation of a monitoring plan, it is often desirable to revisit a number of elements of the plan to continuously refine and update the monitoring activities. In addition, external factors such as changes in monitoring technology, analytical methods, and regulations will often impinge on the design and execution of the monitoring. For these reasons, routine (e.g., annual) reviews of the results and methods should be incorporated into the monitoring plan. However, if trend detection is one component of the plan, care should be exercised in making changes to the program that might compromise the integrity of the data and the ability to use earlier data to infer statistically significant changes in water quality.

### 1.2.5  Other Uses of Resulting Data

Surface water quality data can also be used to support process-based modeling studies using a watershed model such as MAGIC or the Photosynthesis and Evapotranspiration–Biogeochemistry (PnET-BGC) model. Such models can be used to hindcast preindustrial chemistry to determine whether and to what extent a given lake or stream has acidified since the Industrial Revolution. A second general approach is to conduct future scenario modeling to estimate future changes in water chemistry in response to one or more scenarios of emissions control and deposition. A third modeling approach is simulation of the critical or target loads of atmospheric deposition to protect or restore acid-sensitive or nutrient-sensitive aquatic resources (see box on critical and target loads). All of these modeling approaches require compilation of model input data. These can include, depending on the selected model, data on soil chemistry, water chemistry, estimates of historic and current atmospheric deposition, hydrology, and vegetative characteristics.

---

## CRITICAL AND TARGET LOADS

Modeling of critical or target loads requires that a number of decisions be made prior to initiating the modeling. These decisions determine what resources are to be protected, at what level, and over what time period. Sensitive resources to be protected by a given critical or target load can include fish or other aquatic biota, vegetation, or soil condition. To protect these resources, one or more chemical indicators are typically chosen. Often, ANC is used as the indicator for protecting aquatic resources. In that case, one or more critical ANC levels must be selected (i.e., ANC = 20 or 50 $\mu eq/L$), typically in association with known or suspected dose/response relationships for various sensitive species. Different critical ANC levels are expected to protect different species of aquatic life. Finally, one selects a steady-state approach or specifies the time period over which the sensitive resources are to be protected, or over which the damaged resources are expected to recover. Steady-state critical loads are determined irrespective of time. Dynamic critical loads, or target loads, may be determined for various endpoint years, for example, 2050 or 2100. Each of these various decisions that must be made to simulate critical or target loads has an influence on the resulting model simulated values. A target load can be selected that is higher than the modeled critical load if the objective is to make some limited progress toward reaching the critical load. Conversely, a target load can be selected that is lower than the critical load to ensure that the sensitive ecosystem is fully protected given modeling uncertainty or to attain the targeted threshold chemistry more quickly in the case of resources that have already been damaged.

---

# REFERENCES

Baker, J.P., D.P. Bernard, S.W. Christensen, and M.J. Sale. 1990. *Biological Effects of Changes in Surface Water Acid-Base Chemistry*. State of Science/Technology Report 13. National Acid Precipitation Assessment Program, Washington, DC.

Baker, J.P. and C.L. Schofield. 1982. Aluminum toxicity to fish in acidic waters. *Water Air Soil Pollut.* 18:289–309.

Chen, C.W., S.A. Gherini, N.E. Peters, P.S. Murdoch, R.M. Newton, and R.A. Goldstein. 1984. Hydrologic analyses of acidic and alkaline lakes. *Water Resour. Res.* 20(12):1875–1882.

Cosby, B.J., R.F. Wright, G.M. Hornberger, and J.N. Galloway. 1985. Modelling the effects of acid deposition: estimation of long-term water quality responses in a small forested catchment. *Water Resour. Res.* 21(11):1591–1601.

DeWalle, D.R., R.S. Dinicola, and W.E. Sharpe. 1987. Predicting baseflow alkalinity as an index to episodic stream acidification and fish presence. *Water Resour. Bull.* 23:29–35.

Driscoll, C.T., J.P. Baker, J.J. Bisogni, and C.L. Schofield. 1980. Effect of aluminum speciation on fish in dilute acidified waters. *Nature* 284:161–164.

Driscoll, C.T., G.B. Lawrence, A.J. Bulger, T.J. Butler, C.S. Cronan, C. Eagar, K.F. Lambert, G.E. Likens, J.L. Stoddard, and K.C. Weathers. 2001. Acidic deposition in the northeastern United States: sources and inputs, ecosystem effects, and management strategies. *BioScience* 51(3):180–198.

Eilers, J. 2007. *Guidelines for Monitoring Air Quality Related Values in Lakes and Streams in National Forests*. Draft report to the USDA-Forest Service Air Program, Ft. Collins, CO. MaxDepth Aquatics, Bend, OR.

Eshleman, K.N. 1988. Predicting regional episodic acidification of surface waters using empirical techniques. *Water Resour. Res.* 24:1118–1126.

Haines, T.A. and J.P. Baker. 1986. Evidence of fish population responses to acidification in the eastern United States. *Water Air Soil Pollut.* 31:605–629.

Hall, L.W. 1987. Acidification effects on larval striped bass, *Morone saxatilis*, in Chesapeake Bay tributaries: a review. *Water Air Soil Pollut.* 35:87–96.

Havas, M. 1985. Aluminum bioaccumulation and toxicity to *Daphnia magna* in soft water at low pH. *Can. J. Fish. Aquat. Sci.* 42:1741–1748.

Herlihy, A.T. 1997. Section 9. Final lake activities. In J.R. Baker, D.V. Peck, and D.W. Sutton (Eds.), *Environmental Monitoring and Assessment Program Surface Waters: Field Operations Manual for Lakes*. Report No. EPA/620/R-97/001. US Environmental Protection Agency, Washington, DC, pp. 9-1–9-10.

Klauda, R.J., R.E. Palmer, and M.J. Lenkevich. 1987. Sensitivity of early life stages of blueback herring to moderate acidity and aluminum in soft freshwater. *Estuaries* 10(1):44–53.

Lawrence, G.B., K.M. Roy, B.P. Baldigo, H.A. Simonin, S.B. Capone, J.W. Sutherland, S.A. Nierzwicki-Bauer, and C.W. Boylen. 2008. Chronic and episodic acidification of Adirondack streams from acid rain in 2003–2005. *J. Environ. Qual.* 37:2264–2274. doi:2210.2134/jeq2008.0061.

Likens, G.E., F.H. Bormann, R.S. Pierce, J.S. Eaton, and N.M. Johnson. 1977. *Biogeochemistry of a Forested Ecosystem*. Springer-Verlag, New York.

McComick, J.H. and R.L. Leino. 1999. Factors contributing to first-year recruitment failure of fishes in acidified waters with some implications for environmental research. *Trans. Am. Fish. Soc.* 128:265–277.

Muniz, I.P. and H. Levivestad. 1980. Acidification effects on freshwater fish. In: *Proceedings of the International Conference on the Ecological Impact of Acid Precipitation*. September 1980, Oslo, Norway: SNSF Project.

Omernik, J.M. 1987. Ecoregions of the conterminous United States. *Ann. Assoc. Am. Geogr.* 77:118–125.

Passy, S.I., I. Ciugulea, and G.B. Lawrence. 2006. Diatom diversity in chronically versus episodically acidified Adirondack streams. *Int. Rev. Hydrobiol.* 91:594–608.

Sullivan, T.J. (Ed.). 2012. *USDA Forest Service National Protocols for Sampling Air Pollution-Sensitive Waters*. General Technical Report RMRS-GTR-278WWW. USDA Forest Service, Rocky Mountain Research Station, Ft. Collins, CO.

Sullivan, T.J. and A.T. Herlihy. 2007. *Air Quality Related Values and Development of Associated Protocols for Evaluation of the Effects of Atmospheric Deposition on Aquatic and Terrestrial Resources on Forest Service Lands*. Final report prepared for the USDA Forest Service. E&S Environmental Chemistry, Corvallis, OR.

Turk, J.T. 2001. *Field Guide for Surface Water Sample and Data Collection*. Air Program, USDA Forest Service, Washington, DC.

US Environmental Protection Agency. 2000a. *Nutrient Criteria Technical Guidance Manual: Lakes and Reservoirs*. EPA-822-B00-001. US Environmental Protection Agency, Washington, DC.

US Environmental Protection Agency. 2000b. *Nutrient Criteria Technical Guidance Manual: Streams and Rivers*. EPA-822-B00-022. US Environmental Protection Agency, Washington, DC.

US Environmental Protection Agency. 2008. *Integrated Science Assessment for Oxides of Nitrogen and Sulfur—Ecological Criteria*. EPA/600/R-08/082F. National Center for Environmental Assessment, Office of Research and Development, Research Triangle Park, NC.

Webb, J.R., T.J. Sullivan, and B. Jackson. 2004. *Assessment of Atmospheric Deposition Effects on National Forests. Protocols for Collection of Supplemental Stream Water and Soil Composition Data for the MAGIC Model*. Report prepared for USDA Forest Service, Asheville, NC. E&S Environmental Chemistry, Corvallis, OR.

# Chapter 2

# Water Chemistry Field Sampling

## 2.1 BACKGROUND

Water quality studies are much more complex than they first appear. There are so many decisions to make before setting foot in the field. Where will you sample? What will you measure at your sample locations? How will those measurements be made? When will samples be collected? Each of these decision points involves trade-offs that collectively determine what is logistically feasible and what it will cost. Unless you have unlimited time and resources, you will want to think these things through carefully. Your decisions can make or break your study. The notion that you just go out and grab some water samples and then measure what is in them not only is illogical, but also is hugely wasteful of somebody's money and your time. Careful project planning is essential and will reap enormous benefits. We have seen expensive water quality studies that were embarrassingly wasteful and insufficient for addressing the questions at hand. We have also seen modest, clearly targeted studies that yielded useful information. Think about what your questions are, how to best answer them, and how to do that in a way that does not break the bank. The ability to carefully shepherd someone else's resources is a virtue that others will recognize. This is a trait worth cultivating.

In this section, we first address where, what, and when to sample. Decisions are clearly linked to project objectives and research questions. These are issues that you absolutely must master. It is extremely difficult, in conducting your data analyses, to fix a badly designed water quality study. Do it right from the outset.

We then describe field methods, including steps that must be taken before heading into the field, the actual sample collection itself, how to take on-site measurements, and what to do after returning from the sampling sites. This includes measurements of stream flow, various water-sampling approaches, and postsampling processes and sample documentation. Operating principles are presented here. The detailed step-by-step procedures that you will follow are provided in the standard operating procedures (SOPs). Example SOPs are given in Appendices B (stream sampling) and C (lake sampling).

## 2.2   WHERE, WHAT, AND WHEN TO SAMPLE

Data that provide information on the quality of surface water can be used to evaluate the following kinds of issues:

- short-term episodic changes (scale of hours to days) in water quality
- long-term chronic changes (scale of years to decades) in water quality
- types of water quality changes
- likely causes of water quality changes
- longitudinal variation of water quality along streams or depth variation within lakes
- status and extent of chemical and biological condition across populations of lakes and streams
- biological effects of water quality changes

The ability to assess these issues will be limited largely by the extent and intensity of the sampling effort. Three of the most critical aspects of water quality sampling design (cf., Green 1979) are the following:

- concisely and precisely stating the questions to be addressed,
- conducting a preliminary pilot sampling if existing data are not available, and
- replicating sampling in time and in space.

The questions to be addressed will arise from the project objectives; these should be agreed on in advance of field sampling.

Sampling can be conducted at different intensity levels depending on study design, questions to be addressed, and the intended use of the resulting data. The level of intensity will influence decisions about how, where, what, and when to sample and will affect associated quality assurance (QA), laboratory analysis, and data analysis objectives and SOPs. For example, if the objective is to gain a general understanding of the distribution of potentially acid-sensitive streams within a particular wilderness, then a low level of sampling intensity may be perfectly acceptable. This may, for example, entail only a summer survey of specific conductance, pH, or acid-neutralizing capacity (ANC), with no additional measurements. If, however, the objective is to more fully characterize the acid-base chemistry of one or more streams or lakes or to quantify long-term trends in water chemistry, then a higher level of intensity will be required. Sampling that is intended to support regulatory decision making or that will likely be used in permitting or litigation demands perhaps the highest level of intensity. Thus, there is not a one-size-fits-all approach to establishing water-sampling protocols.

The ability to determine the existence of a statistically significant trend in water quality over time is influenced by (1) the magnitude of change that actually occurs in the parameter and water body of interest, (2) the temporal variability that occurs in that water quality parameter and water body, and (3) the number and temporal distribution of samples collected. Thus, to design a monitoring plan to detect the existence of a statistically significant change in lake ANC, one must consider the level of change that one wishes to be able to detect in conjunction with the known or expected temporal variability in ANC in that lake or stream. Prior to initiating a monitoring effort that is intended to evaluate change over time (trends detection), it is helpful to (1) consult with a statistician (or person knowledgeable about statistics) or with the person who will be responsible for the eventual statistical analysis of the resulting monitoring data and (2) conduct a pilot study to determine the temporal variability that occurs in the parameters of interest in that water body (or, at a minimum, in a water body thought to be similar in its chemistry).

The overall data quality objectives (DQOs) for a water quality sampling project are to implement quality control (QC) procedures and requirements for field sampling and laboratory analysis that will provide data that can be used to achieve the program objectives and to follow procedures that will provide data of known quality in terms of precision, accuracy, completeness, representativeness, and comparability. QA/QC issues are covered in detail in the QA/QC protocol section of this book. It is important to note, however, that certain aspects of the QA/QC protocols that are adopted for a particular study will affect choices that need to be made in designing the sampling program for that study. In particular, it is important to determine, in advance of initiating fieldwork, what the DQOs will be with respect to the selected targets for analytical detection limits, precision, accuracy, and completeness. In addition, decisions need to be made concerning how many, and which ones, of the field samples to be collected will be replicated in the field and whether there will be field blanks carried into the field. Note that some sampling programs replicate all samples in the field.

## 2.2.1   Where to Sample

### 2.2.1.1   Selection of Sampling Locations

Selection of sites for water quality sampling should be based on systematic and documented criteria. One of the most important criteria is having a well-defined population of interest. The criteria should be chosen with consideration of watershed factors. These can include representation of bedrock, soil type, geographic distribution of surface waters, presumed sensitivity to stressors of interest, elevation, watershed area, lake area or stream order, site accessibility, and avoidance of watersheds with other elements of human disturbance that might influence surface water composition. Approximate sampling site locations can be identified on a preliminary basis from examination of available mapped data prior to initial sampling trips, with specific site selection and further documentation developed in association with sampling.

#### 2.2.1.1.1   Features of Landscape Associated with Acid Sensitivity of Surface Waters

It is not possible to define *a priori* the features of the landscape that will be most closely associated with surface water acid nutrient or toxin sensitivity nationwide. Such relationships are highly variable across the landscape and should be expected to vary from region to region, sometimes from site to site. Nevertheless, for a given region or study area, it is often possible to identify certain landscape features that correlate with sensitivity or effects. Most commonly, these include such features as lithology, elevation, watershed area, and watershed slope. Sometimes vegetation type (e.g., coniferous forest, alpine and subalpine vegetation), soil type, or one or more regional soil variables (e.g., pH, depth, percentage clay fraction) may also be helpful. Ecoregion designations incorporate many of these variables and can also be useful. In general, the most acid- and nutrient-sensitive lakes and streams are expected to occur under the following conditions:

- bedrock that is *not* basaltic and does *not* contain appreciable amounts of carbonate
- relatively high elevation
- steep terrain
- small watershed
- thin soils
- low soil clay content and low soil pH
- flashy hydrology

In the southeastern United States, south of the line of the most recent glaciation, acid-sensitive streams are often associated with siliciclastic bedrock lithology (cf., Sullivan et al. 2007). In glaciated regions, the presence of varying amounts of glacial till can obfuscate relationships between lithology and surface water chemistry. Sensitive lakes and streams often, but not always, occur at relatively high elevation, on steep slopes, in relatively small watersheds. Knowledge of such relationships, especially if that knowledge is regionally specific, can aid in selection of surface waters for inclusion in inventory and monitoring programs.

## 2.2.1.1.2  *Random versus Nonrandom Site Selection*

One of the most important considerations in site selection is the determination of whether the sampling sites will be selected using a randomized sampling design. Streams or lakes should be randomly selected for sampling if the goal is to characterize populations of surface waters for a defined area too big or impractical to census. This enables the statistics obtained for the sampled waters to be applied to the full population of waters in the designated sample frame within the area. Generally, for a statistically based survey of surface waters, some form of stratified random sampling will be used because this approach allows the sample population to be stratified such that streams or lakes that are of greatest interest can be included in amounts that are disproportionate to their frequency of occurrence in nature. Such a stratified random-sampling process preserves the ability to make population-level extrapolations while maximizing the collection of data for the sites of greatest interest. For example, when it is known that landscape properties such as bedrock or land cover account for spatial variation in surface water quality or susceptibility to degradation, randomized selection and sampling of sites within strata defined by influential landscape properties may allow multiple subpopulation-level extrapolations that collectively provide more information about surface waters in a region than nonstratified randomized sampling. A carefully targeted and stratified random sampling does not necessarily have to entail a large and expensive sampling program. Random surveys of aquatic resources conducted by the Environmental Protection Agency (EPA) have often been large efforts that sampled hundreds to more than a thousand water bodies. These have included the Wadeable Stream Survey (WSS), National Lake Assessment (NLA), National Surface Water Survey (NSWS), and various surveys conducted as part of the Environmental Monitoring and Assessment Program (EMAP). Nevertheless, smaller surveys could also be conducted using a random-sampling structure, thereby allowing extrapolation to a population of waters of particular interest.

If all streams or lakes are included for potential sampling, accessibility may complicate a totally randomized sampling design. This is particularly relevant in remote areas with poor access. Remote sites may require extended periods of time to reach, which lengthens the period over which the survey is conducted and may introduce complications regarding sample holding times and costs. This can also be problematic because environmental sampling conditions (e.g., stream flow) may vary during the survey if some of the sites take several days to access. This can be important because data collected from surface waters sampled during low-flow conditions are generally not comparable to those determined during high-flow conditions.

Note that stratification can be performed on more than one variable or characteristic. For example, within a randomized sampling design, candidate sampling sites can be stratified according to accessibility. This can help ensure that few sites will be included that are remote and difficult to access for sampling. However, use of this approach can result in lower precision in quantifying conditions of remote sites.

Selection of streams or lakes for sampling may be nonrandom if information is needed on specific waters or watersheds rather than on a population of waters or watersheds. Nonrandom sampling may therefore be perfectly acceptable for certain studies. Nevertheless, one should carefully consider that the gains realized in ease of sampling or availability of data collected previously in a nonrandom study must be weighed against the loss of the ability to quantitatively extrapolate directly from the sampling sites to the population of interest.

For nonrandom sampling, especially if the lake or stream is intended to be part of a long-term monitoring (LTM) effort, it may be desirable to select water bodies that exhibit particular characteristics. For example, it makes little sense to spend many years monitoring a body of water that is not acid sensitive if the objective is to evaluate acidification response. Thus, one might purposely select one or more highly sensitive sites for monitoring or for detailed study. Similarly, it would not be logical to focus a study of atmospheric nutrient N enrichment on a surface water that is P limited. For certain studies, it is logical to select a site (or sites) highly sensitive to the stressor in question. Nevertheless, it can be difficult to interpret the results of such studies without first determining where the studied sites fall within the distribution of site sensitivities across the study area or across the region. Such distributions of regional site characteristics can sometimes be provided by various statistically based large synoptic sampling programs, such as EPA's WSS, National Stream Survey, National Lake Survey, or EMAP. Statistically based survey data can be used to aid in selection of sites for LTM. LTM sites may be chosen at random from among randomly selected survey sites so that the resulting monitoring data will be representative of the entire population of interest. This approach was taken in EPA's Temporally Integrated Monitoring of Ecosystems (TIME) project (Kahl et al. 2004).

Thus, connection of survey or monitoring sites to the broader regional population of lakes or streams is almost always highly desired. This connection allows one to extrapolate (whether statistically or merely semiquantitatively) results to more bodies of water than just the ones sampled. Ideally, study sites should be statistically selected. If this is not possible, it may be possible to express the results for a given study site relative to the broader population by quantifying its chemistry relative to the population of lake or stream chemistry determined in one of the larger regional or national surveys, such as those conducted by EPA. Alternatively, the feasibility of conducting a synoptic survey targeted to the specific forest or region should

---

### NUTRIENT STATUS

A rough evaluation of the nutrient status of a lake or stream can be made on the basis of the molar ratio of total N to total P in solution. This determination was formerly based on the Redfield N:P ratio equal to 16. More recent compilations of experimental data (cf. Elser et al. 2009) suggest a cutoff near 44 for N- versus P-limited freshwater lakes. If the ratio is above about 44, the water body is presumed to be P limited, and further addition of N would not be expected to have a large effect on primary productivity. If the ratio is below about 44, the water body is presumed to be N limited and therefore may be sensitive to nutrient enrichment effects from N addition. Such a rough evaluation should be based on multiple samples (at least 10) collected across multiple seasons (ideally spring through fall), as the nutrient status can change with season or with short-term changes in flow or other conditions. Because this is an area of active research, such interpretation should be considered uncertain and subject to change.

A more complete evaluation of nutrient status should be based on laboratory, and perhaps *in situ*, nutrient addition experiments. This might add N and P, individually and combined, to laboratory flasks or *in situ* enclosures, with measurement of nutrient concentrations and chlorophyll *a* or some other measure of primary productivity.

---

be considered. Such a survey could range, depending on resource availability, from a sampling of a few variables to a study of full water chemistry. A screening survey to identify candidate sites for further study could be focused mainly on such parameters as specific conductance and perhaps field pH, with the possibility of full laboratory chemistry for only a subset of the sites.

#### 2.2.1.1.3 Candidates for Inclusion

Prior to conducting either a random or a nonrandom survey of lakes or streams, it is important to determine what kinds (classes) of lakes or streams should be included as candidates for sampling. Preselection of classes of water bodies to include or exclude may change the population frame in statistically based sampling or simply change the waters that are candidates for selection in a nonrandom design. Candidate lakes might be restricted by hydrologic type (to drainage lakes or seepage lakes, for example), lake size, topographic position, ease of access, or depth. Candidate streams might be restricted to

---

### SEEPAGE LAKES

Seepage lakes are lakes that do not have either inlet or outlet streams. There are two general types. Perched seepage lakes are raised above the surrounding terrain, often by buildup of organic deposits; they are often precipitation dominated in their hydrology. Flow-through seepage lakes receive considerable groundwater inputs and generally have higher ANC and silicon (Si) concentrations than do perched seepage lakes.

---

certain stream orders[*] or otherwise constrained according to watershed area, elevation, presence/absence of fish, or presence or absence of watershed disturbance. For nonrandom surveys, intended to identify and characterize the most acid-sensitive surface waters in a specific region, forest, or wilderness, we typically recommend particular focus on the following types of lakes and streams:

- perched seepage lakes
- small (less than about 50 to 100 ha) drainage lakes occupying relatively high landscape position and having average depth greater than about 1 m[†]
- low-order streams (first through third order)

In some cases, a systematic approach to preselection of sampling sites may reduce the number of candidate sites to such a degree that all or most of the high-interest candidate sites can be sampled.

## 2.2.1.2  Selection of Specific Sampling Sites

The sampling site in a lake is generally selected on the basis of logistical considerations. The preferred site is the deepest portion of the lake, but this requires use of a boat, raft, or float tube. If it is not possible to sample at the deepest portion of the lake, then an alternative site can be the lake outlet or (least desirable of the three) a shoreline location.

---

[*] Stream order refers to a system of classifying streams based on their branching pattern. The smallest-headwater streams are first order. When two first-order streams come together, they form a second-order stream. As more first-order streams flow into the second-order stream, its order is not affected; it is still second order. When two second-order streams combine, it becomes third order. The process continues to progressively higher orders. The scale of the mapped data used to designate stream order has influence on the classification. Most acid-sensitive streams tend to be relatively low order (often first through third order at 1:100,000 scale or first through fourth order at higher resolution).

[†] Lakes that are less than about 1 m deep grade into wetlands. Some studies of lakes only include those deeper than 1 m.

A number of factors need to be considered in selecting the specific sampling point in each chosen stream. Any sampling point along a stream will be affected by features of the upstream watershed (Table 2.1). Specific information on the features of the watershed is therefore necessary for determining if the stream is appropriate for sampling for a particular air quality-related value (AQRV), and if so, where on the reach the sampling point should be located. Features that affect water quality include impoundments (ponds, lakes, and reservoirs), wetlands, tributary junctions, distinct changes in slope that alter soils and subsurface hydrology, changes in soil and bedrock type, changes in vegetation, and groundwater discharges (springs). The upstream drainage should also be assessed to determine if disturbances such as fire, mining, or logging have occurred or if the stream has been influenced by erosion from the stream bank or from adjacent roads or land disturbance. In addition, the presence of other human activity in the watershed, such as agriculture or residential development, may affect downstream water quality.

Proximity to trails or roads can be considered in selecting a sampling location. Roads and trails provide accessibility, and a bridge can be used for sampling larger streams and taking flow measurements. If there is an existing stream gauge in the area of interest, colocating the site for stream water sampling with the gauge will provide flow data that would be valuable for interpreting the chemistry data.

As described, the general sampling location can, and should, be prespecified in advance of sending sampling personnel into the field. However, the precise sampling location can be selected by the field personnel when a site is first sampled within a nonrandom survey or monitoring program. Random stream sampling requires that the crews sample at the specified random-sampling point; if that is not possible, then the site is classified as *Not Sampled* and the portion of the population that it represents is categorized as *Not Assessed*.

### TABLE 2.1   LANDSCAPE FEATURES THAT CAN AFFECT WATER QUALITY

Impoundment structure

Wetland

Tributary stream junction

Dramatic change in slope

Abrupt change in vegetation, soil type, bedrock type

Groundwater discharge (spring)

Upslope disturbance (fire, mining, heavily used camping area or trail, logging, windthrow)

Upslope human activity (agriculture, residential development, road building)

Any subsequent sampling of a given site should rely on global positioning system (GPS) coordinates, site photographs, detailed maps, and written
description of the site location to return to precisely the same location each
time that site is sampled. Where allowed, if a site is intended to be sampled
repeatedly, the placement and documentation of uniquely numbered metal
tags at the base of a tree or on a rock adjacent to the sampling site can provide
confirmation of site location.

The field crew should follow these guidelines in selecting new nonrandom
sites for repeated long-term sampling:

- The best point to sample a lake will be the deepest part of the lake. This
  requires use of a boat, raft, or float tube. If it is not possible to collect such an
  open-water sample, the next-best option is to sample the largest flowing
  outlet from the lake; the outlet should be sampled, using stream-sampling
  procedures, as close to the lake as is practical. The third, and least-
  desirable option, is to sample the lake from the shoreline. Shoreline sampling should be conducted, if possible, from a large rock or by wading a
  short distance from the shore. Care must be taken to avoid disturbing the
  sediment in any way that could affect the quality of the sampled water.
  Proximity to logs and aquatic vegetation should be avoided. If possible,
  use wind currents to advantage by collecting the sample from an area
  that receives wind-driven surface water movement from the larger lake.
- The best point to sample a stream will be where the water is flowing fast or falling, where there are no eddies, and where the depth
  is at least 15 cm (6 in.). Ideally, the sampling point is one that can be
  reached during most flow conditions while kneeling on the stream
  bank or on stable rocks downstream from the sampling point. Where
  possible, sites should be selected that allow the sampler to avoid standing or stepping in the water to reach the sampling point and to avoid
  any disturbance of the streambed upstream from the sampling location. Ideally, sites should be selected that allow the sampler to reach
  upstream to collect the sample, well upstream of his or her immediate
  location and well upstream of any location that has been disturbed.
- Stream sample sites should be readily identifiable by reference to semi-
  permanent landmarks, such as confluence points of major tributaries,
  well-marked boundary lines, and stream crossings by permanent roads
  or well-marked trails if they occur in proximity to the selected site.
- Stream sample sites should be selected to avoid direct runoff from
  roads and trails, as well as unmixed flow from tributaries, unless the
  goals of the sampling include those conditions. This will be achieved
  for most small streams by selecting sampling sites at least 50 m above
  road or trail crossings or 50 m above or below inflowing tributaries.

## 2.2.1.3   Establishing and Locating Sampling Sites

For sites that are or will be subject to periodic or routine monitoring, a site information folder or report should be established for each site and provided to the field crew in advance of each sample collection occasion. The site information folder should contain the following:

- driving and site access directions
- maps, including US Geological Survey (USGS) 1:24,000 quadrangle maps and site maps
- estimated travel time from the base location to the sampling site
- overnight lodging or camping information
- local contact personnel (if applicable)
- data collection forms
- permission letters for access (if needed)
- site coordinates and elevation
- site-tag numbers, where applicable (for LTM sites that are marked with a tree tag), and locations (not allowed in wilderness)
- site photographs
- other relevant information

Maps provided in the site information folder may also include forest recreation maps to help navigate to the area. Maps generated using geographic information systems (GISs) could also be included to show where the project manager has selected potential sites to sample, spatial patterns in the distribution of vegetation types or other landscape properties (e.g., soil or bedrock distribution), or locations where sites were sampled in a previous study.

Lake or Stream Water-Sampling Record data sheets will serve for documenting site information, sample locations, and field measurements. These forms should be printed on waterproof paper. Copies of these data sheets are provided in Appendix D.

In addition to the material described for inclusion in site documentation folders, site documentation materials can include

- uniquely numbered aluminum tags (where allowed) for sites planned to be sampled repeatedly (i.e., monitoring sites) or for replacement of missing tags at previously established sites,
- nails and a small hammer for tag placement,
- flagging tape,
- a camera with a date/time stamp for site photographs,
- a GPS unit for determination of geographic coordinates (in decimal degrees), and
- waterproof pens for completing forms in the field.

Depending on the objectives of the field data collection, field crews may be collecting water samples as part of a synoptic survey, or they may be repeating sampling at the same locations in a monitoring effort to examine changes in water chemistry over time. The extent of documentation required by the field crew in the field will depend on whether the site is new or previously established.

## 2.2.2   What to Measure

This book focuses on protocols for water quality sampling to quantify the effects of atmospheric deposition on aquatic ecosystems. The atmospheric deposition constituents of concern described here are S, N, and toxics. It is possible to offer straightforward guidance regarding sampling constituents associated with characterization or monitoring of the acidification and nutrient enrichment effects of S and N deposition. The constituents to monitor for studies of the effects of toxics are more variable depending on the objectives of the particular study and are therefore less subject to generalization.

The primary water quality variables to be sampled can include physical, chemical, or biological attributes. Choice of variables depends on the potential environmental risks, logistical issues associated with sampling for these parameters, and costs.

A water quality survey or monitoring for the purpose of evaluating responses to atmospheric deposition can involve any number of parameters. The choice of parameters should clearly relate to the water quality concerns and should be measurable in a routine sampling program. The challenge is to select those parameters that are most important with respect to revealing key features of ecological integrity and that can be determined in a relatively straightforward and cost-effective fashion. For some studies, samples are needed at sufficient frequency and temporal resolution that they allow appropriate characterization or statistical trend detection in the future.

The choice of what to measure will depend in large part on the type of study:

- acidification
- eutrophication
- bioaccumulation or toxicity

Parameters to include in each of these kinds of studies are discussed next.

### 2.2.2.1   Acidification Studies

Atmospheric inputs of both S and N can cause acidification of soil, soil water, and fresh drainage water (lakes, streams). In most regions of the United States that have experienced acidification impacts from air pollution, those impacts have mainly been caused by S deposition. There are also, however, some regions, especially in the western United States, where resources are more threatened by N inputs than by S inputs. This is at least partially because of the very low levels of S deposition received at many western locations. There are also regions (portions of the Northeast, West Virginia, high elevations in North Carolina and Tennessee) where both atmospheric S and N contribute substantially to the observed acidification in some lakes and streams.

Acidification from S and N deposition can have several important chemical and biological effects. In particular, there are changes in the acid-base status of surface and soil water that can cause short-term or long-term toxicity to aquatic or terrestrial biota.

Watershed processes control the extent of ANC contribution from soils to waters as drainage water moves through terrestrial systems. These processes regulate the extent to which drainage waters will be acidified in response to acidic deposition. Of particular importance is the concentration of acid anions in solution, including sulfate ($SO_4^{2-}$), $NO_3^-$, and organic acid anions. Naturally occurring organic acid anions, produced in upper horizons of acid-sensitive soils, normally are removed from solution as drainage water percolates into the deeper mineral soil horizons. In some regions, organic acids can dominate the acid-base chemistry of a lake or stream (as indicated by color and dissolved organic carbon [DOC] concentration) because of the occurrence of hydrologically connected wetlands. Organic acids derived from wetlands, although they acidify a lake or stream, also serve as buffers against further pH depression from acidic deposition. Acidic atmospheric deposition allows natural soil acidification, anion mobility, and cation-leaching processes to occur at greater depths in the soil profile, allowing water that is rich in $SO_4^{2-}$ or $NO_3^-$ to flow from mineral soil horizons into drainage waters. If these anions are charge balanced by $H^+$ or $Al_i$ cations, the water will have low pH and could be toxic to aquatic biota. If they are charge balanced by base cations (BCs), the pH of the water will be higher, but the BC reserves of the soil can become depleted over time.

Nitrate and ammonium ($NH_4^+$) have the potential to acidify surface waters. However, N is also a limiting nutrient for plant and microbial growth in most terrestrial, and some aquatic, ecosystems. Therefore, atmospheric N deposition can contribute to increased productivity, eutrophication, and N saturation in some surface waters. This appears to most frequently be the case in estuaries and near-coastal marine waters and in freshwaters in remote locations where historic atmospheric N deposition has been low.

High concentrations of lake or stream water $NO_3^-$ may be indicative of ecosystem N saturation, reflecting a condition in which the supply of N exceeds the biological demand. Nitrogen saturation has been found at a variety of locations throughout the United States. These have included the San Bernardino and San Gabriel Mountains within the Los Angeles Air Basin (Fenn et al. 1996); the Front

Range of Colorado (Baron et al. 1994, Williams et al. 1996); the Allegheny Mountains of West Virginia (Gilliam et al. 1996); the Catskill Mountains of New York (Murdoch and Stoddard 1992, Stoddard 1994); and the Great Smoky Mountains in Tennessee (Cook et al. 1994).

The mobility of $SO_4^{2-}$ is an important factor governing the extent to which S deposition contributes to soil and water acidification, BC depletion, and Al mobilization, each of which can harm sensitive ecosystems. Sulfur deposition moves through watershed soils and into surface waters as $SO_4^{2-}$. Sulfate is the most important anion contributed by acidic deposition in most, but not all, parts of the United States. In some regions (including the glaciated Northeast, Upper Midwest, and West), much of the deposited S moves readily through soils into streams and lakes. Thus, $SO_4^{2-}$ has been classified as a mobile anion (Seip 1980). However, $SO_4^{2-}$ is less mobile in areas having older, more weathered and nonglaciated soils, most notably the southeastern United States.

One of the most important effects of acidic deposition on watersheds has been increased mobilization of Al from soils to surface waters (Cronan and Schofield 1979). Aluminum, which occurs naturally in soils, has a pH-dependent solubility in water. Solubility increases dramatically at pH values below about 5.5. Aluminum concentrations in acidified drainage waters are often an order of magnitude higher than in circumneutral waters. Effects of Al mobilization to surface and soil waters include toxicity to aquatic biota (Driscoll et al. 1980, Muniz and Levivestad 1980, Schofield and Trojnar 1980, Baker and Schofield 1982); toxicity to terrestrial vegetation (Ulrich et al. 1980); alterations in nutrient cycling (Dickson 1978, Eriksson 1981); and pH buffering effects (Driscoll and Bisogni 1984). Inorganic monomeric Al concentrations often reach potentially toxic concentrations (> about 2 μM) in surface drainage waters having pH less than about 5.5.

There can be substantial leeway in terms of selection of parameters to measure in a field study of surface water acidification. Analytical costs must be weighed against the value contributed by each constituent that one may choose to analyze in the laboratory. In general, we recommend that the parameters in Table 2.2 be considered the core for inclusion in the suite of analytes to be measured in any study of surface water acid-base chemistry. When budgets allow, all of the parameters listed in Table 2.2 that are designated as having high importance should be included in the list of analytes, plus DOC. If pH is below about 5.5, we also recommend analyzing for total monomeric Al ($Al_m$) and $Al_o$. The concentration of the potentially toxic $Al_i$ is then obtained by subtracting $Al_o$ from $Al_m$. Although DOC should be measured in all acid-base chemistry studies, color could be substituted as an inexpensive alternative if necessary. Measurement of dissolved inorganic carbon (DIC) should be considered optional; this measurement can be used in estimating $HCO_3^-$ concentration, which is important as part of

**TABLE 2.2   PARAMETERS TO CONSIDER FOR POSSIBLE INCLUSION IN SURFACE WATER ACIDIFICATION STUDIES**

| Parameter | Preferred Unit | Importance | Rationale |
|---|---|---|---|
| ANC | µeq/L | High | ANC is the master acid-base chemistry variable in aquatic systems |
| pH | — | High | Biota respond strongly to pH |
| $SO_4^{2-}$ | µeq/L | High | Usually the major acid anion from atmospheric deposition |
| $NO_3^-$ | µeq/L | High | Sometimes an important acid anion |
| $Ca^{2+}$ | µeq/L | High | Usually the major base cation |
| $Mg^{2+}$ | µeq/L | Moderate | Usually an important base cation |
| $K^+$ | µeq/L | Moderate | Base cation, usually in low concentrations |
| $Na^+$ | µeq/L | Moderate | Indicator of road salt contamination, geological sources, or sea salt inputs |
| $Cl^-$ | µeq/L | Moderate | Indicator of road salt contamination, geological sources, or sea salt inputs |
| $NH_4^+$ | µeq/L | Moderate | Potential indicator of agricultural influence or anaerobic conditions |
| Specific Cond. | µS/cm | Moderate | Useful in QA evaluation of internal data consistency; potential general screening variable to identify waters with low ionic strength |
| DOC | µM | Variable | Indicator of organic acidity |
| $Al_m$ | µM | Variable | Used with $Al_o$ to estimate potentially toxic $Al_i$ |
| $Al_o$ | µM | Variable | Used with $Al_m$ to estimate potentially toxic $Al_i$ |
| Si | µM | Variable | Potential indicator of lake hydrologic type and groundwater inputs; may explain some patterns in diatom presence and abundance |
| DIC | µM | Low | Used to estimate $HCO_3^-$ concentration |
| Total dissolved F | µM | Low | Used for $Al_i$ speciation |

cond. = conductance

the charge balance. Some studies might choose to analyze silicon (Si) or total fluoride (F), but these are often not needed for standard acid-base chemistry assessment. The concentration of Si can be useful in evaluating the extent of groundwater influence on surface water chemistry and in discriminating between perched and flow-through seepage lakes. It can also provide useful information in interpreting diatom data because Si can be limiting to diatom growth in some cases. Measurement of total dissolved F (fluorine) is needed to calculate the speciation of $Al_i$ into various components, such as $Al(OH)^{2+}$, $Al(OH)_2^+$, $AlF^{2+}$, $Al(F)_2^+$, $Al^{3+}$, and so on. This can be important because the Al-F species are thought to be less toxic to aquatic biota than the Al-hydroxide species and $Al^{3+}$.

Important parts of the water chemistry QA/QC evaluation can include determination of the charge balance and comparison between measured and calculated conductivity, sum of anions and sum of cations, and titrated and calculated ANC. Charge balance calculations can also be used to determine the charge density (organic anion concentration per mole of DOC) of DOC in surface waters. To permit these QA/QC checks to be conducted, all parameters listed in Table 2.2 are required except Si. Thus, the full list of parameters should be analyzed if funding permits. It is possible to perform these evaluations without a measurement of total dissolved F if one is willing to make certain assumptions about the $Al_i$ speciation.

## 2.2.2.2 Eutrophication Studies

Eutrophication, or nutrient enrichment, is a potential consequence of N deposition to aquatic ecosystems that are N limited. However, many freshwater ecosystems are P limited and therefore would not be expected to increase primary productivity in response to increased atmospheric inputs of N. Nevertheless, there are many examples of freshwaters that appear to be N limited or N and P colimited (e.g., Baron 2006, Elser et al. 2009). In such aquatic systems, atmospheric inputs of N would be expected to increase productivity or alter biological communities such as phytoplankton.

Atmospheric deposition of N may increase in the future in remote areas that are situated downwind from centers of agricultural or human population growth. Surface waters in such areas can be N limited. As a consequence, N additions might contribute to nutrient enrichment, including changes in algal species distribution and abundance. In particular, high-elevation areas in the Sierra Nevada and Rocky Mountains (and perhaps portions of the Cascade Mountains) are susceptible to such increases in nutrient N deposition (Fenn et al. 2003, Sickman et al. 2003b). In some areas, atmospheric N deposition has been linked with eutrophication of high-elevation lakes (cf., Melack et al. 1989, Sickman et al. 2003a).

Estuaries and near-coastal marine ecosystems are also susceptible to nutrient enrichment, especially from N. This is because estuarine and marine waters tend to be N limited. Land clearing, agricultural land uses, sewage treatment discharge, and atmospheric deposition can all result in high loadings of N to coastal zones. Excessive N inputs can contribute to a range of impacts, including enhanced algal blooms, decreased distribution of seagrass habitat, and decreased dissolved oxygen (DO) concentration (Valiela et al. 1992, Nixon 1995, Borum 1996, Bricker et al. 1999, Kopp and Neckles 2004). Because of human population growth and urban development in coastal areas, there is substantial potential for increased N loading to coastal ecosystems. Atmospheric deposition of N contributes to that load but is generally not the major source of estuarine N. AQRVs for protection of estuarine ecological conditions are beyond the scope of this book. Recommendations for monitoring estuaries and other coastal areas are therefore not addressed.

There is no clear-cut selection of chemical parameters to include in a study of potential eutrophication of lake or stream water. A variety of measurements can be useful (Table 2.3). In general, measures of N, P, and chlorophyll $a$ are of greatest importance. We recommend, at a minimum, that water samples be analyzed for total N, $NO_3^-$, $NH_4^+$, total P, soluble reactive phosphorus (SRP), and chlorophyll $a$. In addition, dissolved organic N (DON) may be of interest. The measurement of SRP is intended to reflect the forms of P in surface waters that are most readily available to aquatic biota. Nevertheless, P forms are to some extent interchangeable within the water column and stream/lake sediment. Therefore, measured total P (which includes both soluble and particulate forms) is also of interest in evaluating potential nutrient limitation and growth responses. In general, $NH_4^+$ and $NO_3^-$ are considered to be biologically available forms of N. Nevertheless, DON may be converted to $NH_4^+$ and $NO_3^-$ or used directly by some primary producers. Therefore, measured total N is also of interest. Additional physicochemical parameters that can be useful in evaluation of nutrient status include iron (and perhaps other metals); Si (lakes only); DO; total suspended solids (TSS); turbidity; $Ca^{2+}$; total Al; and Secchi depth (a physical, rather than a chemical, measurement). Iron, $Ca^{2+}$, and Al can bind to P and influence its cycling between sediment and water and also its bioavailability. Silicon can be limiting or colimiting, along with P and N, to diatom productivity. It can also provide information regarding groundwater inflow to a lake. High productivity in response to nutrient enrichment can lead to reduction in DO as primary producers die and decay, consuming oxygen ($O_2$) through microbial respiration. This effect is generally associated with rather extreme eutrophication, well above the levels that might be expected to occur in response to atmospheric deposition inputs to freshwaters in the United States. The TSS concentration is useful because eroded sediments, especially the smaller clay-size particles, can be relatively enriched in adsorbed P, depending on local geology and land use. Thus, eroded sediments contribute to the total P in surface waters, especially in streams during high-flow periods. At locations where the

**TABLE 2.3    PARAMETERS TO CONSIDER FOR POSSIBLE INCLUSION IN STUDIES OF ATMOSPHERIC NUTRIENT N ENRICHMENT OF FRESHWATERS**

| Parameter | Preferred Unit | Importance | Rationale |
|---|---|---|---|
| Total N | μM | High | Reflects all forms of N in the system |
| $NO_3^-$ | μM | High | Biologically available form of N |
| $NH_4^+$ | μM | High | Biologically available form of N |
| Dissolved organic N (DON) | μM | Moderate | Potentially available form of N |
| Total P | μM | High | Reflects all forms of P in the system |
| Soluble reactive P (SRP) | μM | High | Biologically available P |
| Chlorophyll *a* | μg/L | High | Reflects primary productivity |
| Fe | μM | Variable | May bind with P, influencing its bioavailability and transport |
| Total Al | μM | Variable | May bind with P, influencing its bioavailability and transport |
| $Ca^{2+}$ | μM | Variable | May bind with P, influencing its bioavailability and transport |
| Si | μM | Variable | May be limiting to diatoms under some conditions |
| Dissolved oxygen (DO) | mg/L | Variable | May decrease to biologically stressful levels under extreme conditions of nutrient inputs (under most conditions of atmospheric nutrient deposition, decreased DO is not an important issue) |
| Total suspended solids (TSS) | mg/L | Variable | May be an erosional source of P to streams |
| Turbidity | Standard units | Variable | May be used to estimate TSS |
| Secchi depth | m | Variable | Can reflect algal abundance in lakes |

geology contains substantial P, this effect can be quantitatively important. It is therefore helpful to evaluate local geology and human activities throughout the entire watershed, especially land use actions that contribute substantial erosion to surface waters. Turbidity can sometimes be used (along with an appropriate training set that includes simultaneous measures of TSS and turbidity) to estimate TSS in stream water. Therefore, it is possible to rely on routine measurement of turbidity and infrequent measurement of TSS (a more expensive laboratory analysis), along with an empirical relationship between the two in order to estimate TSS for all sample occasions. Depending on the stream, this approach can yield relationships that are more or less robust. Secchi depth provides an indication of relative algal density in lake water. It can therefore provide a good index of algal production.

### 2.2.2.3 Bioaccumulation and Toxicity Studies

Atmospheric deposition can contribute to toxicity of surface water in several ways. The atmospheric pollutants of greatest concern with respect to toxicity, in addition to $H^+$ and $Al_i$ associated with acidification, are primarily pesticides, mercury (Hg), and other trace metals.

#### 2.2.2.3.1 Pesticides and Other Toxics

Pesticides and other toxics can be air deposited, and some can bioaccumulate in predator species. The degree of bioaccumulation is generally a function of the age of the organism and its position in the food web. In general, older individuals at the top of the food web have bioaccumulated more toxic materials than have younger individuals nearer to the bottom of the food web.

Pesticides applied to agricultural crops can become volatized or suspended in the atmosphere with dust particles and eventually be transported with prevailing winds to remote areas. For example, organophosphate pesticides have been detected in precipitation at elevations up to 1920 m in Sequoia National Park in California (Zabik and Seiber 1993) and measured in plant foliage across a range of elevations (Aston and Seiber 1997). The effects of atmospheric deposition of pesticides at remote locations are poorly known. However, there is particular concern that fungicide deposition could harm sensitive lichen species (McCune et al. 2007).

#### 2.2.2.3.2 Mercury

A variety exists of other toxic chemicals that can be atmospherically deposited, some of which have the potential to bioaccumulate. These include trace metals, polychlorinated biphenyls (PCBs), and some fire-retardant chemicals. The toxin considered to be of greatest concern is usually Hg.

Atmospheric deposition is an important component of Hg cycling and biogeochemistry. Mercury is naturally occurring and is found throughout the environment. Mercury present in fossil fuels is released to the atmosphere during combustion and is subsequently available for long-range atmospheric transport and deposition to Earth's surface. Coal combustion in power plants is a major source of atmospheric Hg. It enters lakes and rivers from atmospheric deposition of the Hg emitted by air pollution sources and from nonpoint sources via erosion and runoff.

Mercury is known to bioaccumulate in aquatic organisms, reaching potentially high concentrations in larger, piscivorous fish and the species that consume them. Such Hg bioaccumulation is an important human health concern, especially among subpopulations of people who consume large quantities of fish, children, and women of childbearing age. Mercury is a toxin that can damage the human brain and nervous system. Human exposure to Hg mostly occurs via consumption of fish and other seafood that has accumulated high concentrations of Hg. Fish consumption advisories for various lakes and rivers have been issued in most states throughout the United States. Mercury can also accumulate in fish-eating wildlife such as loons, river otters, bald eagles, and other piscivores.

Monitoring studies to evaluate the extent to which atmospheric deposition of Hg affects aquatic ecosystems could focus on the concentrations of total or MeHg in water, invertebrates, fish, or piscivorous birds and mammals. Alternatively, methylation rates or bioaccumulation within different environmental compartments might be quantified.

We do not recommend widespread efforts to measure or monitor Hg concentrations in surface waters as part of routine water quality evaluation. Rather, focused studies are recommended in areas where atmospheric deposition of Hg is known or suspected to be high. Such focused studies might begin by investigating Hg concentrations in muscle tissue of a fish biomonitor such as yellow perch (*Perca flavescens*) or large piscivorous fish such as bass. Such data might be more useful than measurement of Hg concentrations in water to make preliminary judgments regarding Hg cycling and toxicity issues. Furthermore, measurement of ambient Hg concentrations in water is technically difficult and requires advanced training of field crews.

## 2.2.3   When to Sample

There is no one answer to the question of when to collect samples of surface water for evaluating potential impacts of atmospheric deposition. The answer depends on the type of study and its specific objectives. A general breakdown of sample timing is given in Table 2.4. It is fairly standard procedure to target sampling to a particular season under base flow conditions. Lakes are often sampled during summer or fall base flow. Streams are often sampled during spring base flow for acidic deposition research and summer for nutrient work.

**TABLE 2.4    SUGGESTED TIMING OF SURFACE WATER SAMPLES FOR EVALUATION OF SENSITIVITY TO AND EFFECTS FROM ATMOSPHERIC DEPOSITION OF ACIDIFYING OR EUTROPHYING SUBSTANCES**

| Type of Study | Suggested Timing of Sample Collection |
|---|---|
| Lake or stream characterization | Index period – at least one sample each year for at least three years, *and* |
| | High-flow period (snowmelt and/or rainstorm) - at least two samples, if possible |
| Synoptic survey | Index period – at least one sample at each site within a relatively narrow time period; avoid high-flow conditions |
| Characterization of episodic chemistry during hydrologic events | High-flow period – at least three samples during each of at least three hydrologic events, including at least one large storm (one year storm or larger) and (if applicable) one substantial snowmelt event |
| Long-term monitoring | **Acceptable Approach**<br>Index period – at least one sample per year within a relatively narrow time period or hydroperiod; avoid high-flow conditions |
| | **Preferred Approach**<br>Index period – at least one sample per season during the open-water seasons, within relatively narrow time periods or hydroperiods |
| Modeling with MAGIC or PnET-BGC model | **Acceptable Approach**<br>Index period – at least one sample |
| | **Preferred Approach**<br>Index period – at least one sample during each of the open-water seasons during at least three years |

*Note:* MAGIC, Model of Acidification of Groundwater in Catchments; PnET-BGC, Photosynthesis and Evapotranspiration–Biogeochemistry.

Streams in the southeastern United States, where snowmelt is not a major hydrologic factor, are often sampled during spring. Samples collected under base flow conditions are often used to evaluate long-term changes in water quality. Short-term changes are more commonly evaluated under episodic conditions influenced by snowmelt or rainstorms. Because of the possibility for either episodic or incremental degradation of water quality in response to atmospheric deposition, it may be important to implement a program for monitoring both short- and long-term changes. For some studies, it may be desirable to avoid sampling during abnormally low- or high-runoff conditions. Other studies may be focused on extreme flow conditions. USGS discharge data (e.g., North Carolina data are found at http://nc.water.usgs.gov/info/h2o. html) can be examined prior to going to the field to evaluate ambient stream flow from stream gauges in the general area of the sample site. This precaution is more important when sampling stream, as opposed to lake, chemistry and when water characterization will be based on a single sample rather than multiple samples collected at different times. It is also important to consider the potential influence of climatological wet/drought cycles on the chemistry of surface waters.

### 2.2.3.1   Lake or Stream Characterization

The water chemistry of lakes or streams is often characterized on the basis of a single sample, collected during the index season (spring or summer for streams; summer or fall for lakes). In general, however, it is preferable, but not necessary, to base surface water characterization, assessment, or modeling on multiple samples (either collected throughout the annual cycle or restricted to the index season) collected over several years.

### 2.2.3.2   Synoptic Survey

The time at which the water sample is collected during a synoptic survey can influence the resulting chemistry and the ways in which the data can be used and interpreted. Stream water chemistry, and to a lesser extent lake water chemistry, can vary diurnally (Burns 1996), seasonally (Lawrence et al. 2004), and annually (Murdoch and Shanley 2006). Stream water chemistry can also vary on an hourly basis in response to changes in flow (Lawrence 2002). Therefore, if the objective is to conduct a synoptic survey to compare measurements among different streams or lakes within the same region (a spatial assessment), all sites would ideally be sampled at the same time, which is seldom possible. Even this approach might not ensure the same sampling conditions if a localized storm was affecting only part of the study region. Nevertheless, approaches can be chosen to minimize the effects of temporal variability when conducting synoptic sampling.

The major causes of temporal variability in water quality are generally associated with flow and climate. Diel changes in pH and metal concentrations caused by patterns of photosynthesis and respiration can also be important. Climate is particularly important in regions where snow accumulation occurs in winter, and water chemistry is affected by snowmelt (Campbell et al. 1995, Lawrence et al. 2004). Seasonal effects can be easily addressed by restricting the sampling to a period that falls within a single season. However, the choice of season can also affect the frequency and magnitude of flow variations. For example, in some regions, summer is the season that typically has the lowest flows and the fewest rain- or snowmelt-driven hydrological events. Scheduling a synoptic survey during a period of stable flows in such regions is therefore more easily accomplished during the summer as compared with other seasons.

The choice of sampling season may necessitate collection of samples when flow variations are relatively frequent and substantial. Extreme chemical conditions are often associated with extreme flows. This is the case with some acidic deposition effects, which tend to be most severe in some regions during the high, fluctuating flows of spring snowmelt or the large rainstorms associated with hurricanes or other major storm systems. Although these are the conditions most difficult to characterize, they are often highly relevant for assessing biological impacts.

Two approaches can be used in synoptic studies to address flow variations within a season, but each requires stream flow gauges within the sampling region to monitor flow conditions during the sampling period. In the first, sampling of each stream is repeated multiple times during the season under a variety of flow conditions. If the mean and distribution of flows at the collection times are similar (not statistically different) among the streams, the mean or median chemical concentration at each site can be used as the representative value for comparing streams. This approach was used successfully in the first large-scale stream survey to assess acidic deposition effects in the United States (Colquhoun et al. 1984).

The second approach involves collection of samples at all sites during a period of time that has limited variation in flow. This requires close monitoring of weather conditions coupled with the ability to initiate or interrupt the sampling on short notice. This approach was successfully used in two snowmelt surveys by Lawrence et al. (2008) in the Adirondack region of New York. Collection of all stream samples was done over 3 days when flows were elevated but stable (higher than 90% of the year in one survey and higher than 84% of the year in a second survey). This approach can be challenging to implement with stream-sampling sites that are difficult to access or far apart.

Neither of these approaches is likely to collect samples during the most extreme conditions. However, the stream gauges used to monitor flow in the region can be used to determine how the flow conditions during the sampling window related to conditions for the overall year (and previous years). If information is available to show that the chemical measurement of interest is statistically related to flow, then it is possible to estimate the measurement for more extreme flow conditions based on this relationship and available flow data. For streams where flow measurements are not available, chemical concentrations for flows higher than those at the time of sampling can be approximated from an index stream where flow and chemical concentrations are monitored on a regular basis throughout the year. Examples of this approach were given by Eshleman (1988) and Lawrence et al. (2008).

### 2.2.3.3 Characterization of Episodic Chemistry during Hydrologic Events

Characterization of episodic chemistry (generally of streams, occasionally also of lakes) during hydrologic events is challenging under the best of circumstances. Given the additional complications of access difficulties for remote sites, often long travel distances, and complications and safety concerns introduced by seasonal snowpack, episodic sampling in backcountry settings is seldom attempted. Nevertheless, such sampling can yield important information to aid in interpretation of surface water acid-base and nutrient chemistry.

Because of the transient and unpredictable nature of hydrologic events, precise timing of sample collection occasions is generally not possible. In particular, it cannot be assumed that such samples necessarily capture the most extreme chemical conditions. For that reason, multiple high-flow samples should be collected, if possible, and they should ideally be distributed across multiple years.

### 2.2.3.4 Long-Term Monitoring

If an LTM program for lakes is intended to represent a particular time of year, interannual variation in lake hydroperiod (periodicity in lake conditions that reflect the changes of the seasons, including water and air temperature, snowmelt, and vegetative development) can introduce substantial variability. This can be especially problematic for high-elevation lakes. The chemistry of such lakes can change gradually or abruptly in response to spring snowmelt, large rainstorms, or fall overturn. Such changes are largely governed by the depth of the snowpack, patterns of rainfall, and temperature. Thus, a program that entails, for example, sampling the first week of July each year, while reducing some aspects of interannual variability, may still yield considerable year-to-year variability as a consequence of interannual differences in snowmelt hydrology. One potential solution is to target sampling to a specific degree day, which is calculated based on maximum and minimum daily temperature (cf. http://pnwpest.org/wy/index.html). Standardization of sampling timing on the basis of degree day can partially adjust for interannual differences in snowmelt and the transition to summer weather. It therefore should eliminate some, but not all, of the variability associated with standardization based on the calendar for certain types of lakes. Such an approach should not, however, ignore the influence of flow conditions.

One can quantify changes over time in the concentration-discharge relationship. Using this approach, one compares differences between the concentration-discharge relationship determined during one period of time (several months to several years) and the same relationship determined during a later period of time of similar length. If, for example, the concentration of stream ANC is higher at a given flow condition this year than it was several years ago, this pattern may suggest that ANC may be increasing at times represented by that flow regime, independent of any changes in flow.

Analysis of trends is most often done on an annual basis using one of several approaches that incorporate seasonal effects (Helsel and Hirsch 1992, Lawrence et al. 2004). These approaches are most effective if multiple samples are collected for each season. Weekly sampling provides a sufficient number of samples to account for within-season variability and is likely to enable a trend to be detected with fewer years of monitoring data than data collected at longer intervals. Annual or quarterly sampling is less expensive than weekly sampling but cannot account for within-season variability and, relative to weekly sampling, can substantially increase the length of time needed to detect a trend (Murdoch and Shanley 2006). However, annual or quarterly, as opposed to weekly, sampling may free up resources to monitor more sites to obtain a better picture of regional patterns. Thus, the intended eventual use of the data is important for making sampling decisions, as is the length of time one is willing to wait before being able to document with statistical certainty that a change has taken place.

Because LTM of surface water chemistry is usually based on sampling at a constant frequency, most samples are typically not collected during high-flow periods. However, long-term trends in stream chemistry may first become apparent during high flows. An approach for separate trend analysis of high, medium, and low flows has been developed (Murdoch and Shanley 2006). This method uses annual or grouped years of data to develop concentration-discharge relationships that enable concentrations to be predicted for various flow conditions throughout the year. An annual value can then be derived for upper-, medium-, or low-flow ranges so that long-term trends can be determined for each specific flow range. This type of approach requires that (1) flow is monitored for the stream site of interest, (2) the solute of interest is statistically related to flow, and (3) sufficient data are available to develop the concentration-discharge relations.

## 2.3  FIELD METHODS

The methods outlined here are appropriate for analysis of stream and lake waters with low ionic-strength and associated with forested and alpine watersheds in lands that are sensitive to acidification, toxicity, or nutrient enrichment impacts from atmospheric deposition. Because stream and lake waters in the most highly sensitive areas can be extremely dilute (and therefore easily contaminated), great care must be taken in all phases of sample collection and analysis to ensure that samples are not contaminated during collection or processing so that data will be of sufficient quality to support the intended assessment purposes. Each of the important aspects of field sampling is discussed here, with an explanation of the reasons why certain steps should be taken, or avoided, in the sampling program. The intent is to provide a general understanding of sampling issues. The specific, step-by-step instructions to the field personnel can be developed by the project manager and documented in the SOPs (see material presented in Appendices B and C). Here, we recommend core attributes and standard sampling design and procedural elements applicable across the country. This can then be used by water quality practitioners to develop the site-specific project plan that includes the SOP that lists the specific steps to be followed by field personnel. The project plan and SOPs may vary from study to study, but the overall principles of investigation remain constant. General SOPs are provided in Appendices B and C. They can be modified, as needed, for a particular study.

## 2.3.1  Pretrip Preparations

A field data sheet should be prepared in advance for each sampling site. Example Lake- and Stream-Water-Sampling Record data sheets are provided in Appendix D. The appropriate data sheet should be completed to the extent possible prior to the sampling trip. For previously established sites, the available site information, site-tag number, and description of the tag tree (where applicable) should be filled out on each form.

Pretrip preparation should include an evaluation of sample holding time issues. It is important to determine, in advance, what parameters will be analyzed in the laboratory and then to check the laboratory protocol to determine the holding time requirements of these measurements. This may have an influence on sample collection scheduling or in-field sample aliquot preservation decisions.

It is also important to check with the laboratory regarding the timing of sample delivery. In general, sampling should not be done late in the week or in advance of a holiday unless arrangements have been made with the laboratory to receive samples at those times.

The most important issues to consider prior to entering the field can vary depending on the type of study. A checklist of important issues to consider in advance of initiating and implementing an inventory or monitoring field effort is provided in Table 2.5. Additional issues to consider prior to initiating an LTM (trends) effort are listed in Table 2.6; Table 2.7 provides a list of issues to consider when conducting a study that will involve critical load or emissions scenario modeling.

### 2.3.1.1  Permits and Access

Atmospheric deposition effects sampling often occurs on public lands, including wilderness areas, national parks, national forests, and state parks and forests. Thus, access permission is generally not required. Nevertheless, motorized access to a trailhead or other access point in close proximity to a sampling location may necessitate crossing private property. This may require acquiring access permission from the landowner. Obtaining such permission can sometimes be a lengthy process.

**TABLE 2.5 KEY ISSUES TO CONSIDER IN INITIATING AND IMPLEMENTING AN INVENTORY OF WATER CHEMISTRY AT ONE OR MORE LAKES OR STREAMS FOR THE PURPOSE OF ASSESSING EFFECTS OF ATMOSPHERIC DEPOSITION**

### Site Selection Issues

1. Is the lake/stream representative of other lakes/streams in the wider regional population? How sensitive to the stressors of interest is it expected to be?

2. Is the sampling site and its upstream drainage basin reasonably free of unwanted or unquantifiable disturbances other than atmospheric deposition (i.e., acid mine drainage; geological S; fertilizer application; livestock influence; riparian, in-channel, or shoreline disturbance)?

3. Is the sample site representative of the body of water being sampled? In other words, is the lake sampling site in the deepest part of the lake or alternatively in the outlet stream? Is the stream sampling site in the thalweg (main region of water flow, usually located toward the middle of the stream), well below the nearest upstream confluence?

### Implementation Issues

4. Have arrangements been made, in advance, with the laboratory and any required permits or property access permissions obtained?

5. Have issues associated with laboratory holding times and length of time needed for field site access and delivery of samples to the laboratory been addressed?

6. Have clean, appropriate size bottles and (if required) syringes been obtained?

7. Have decisions been made about field QA/QC activities, including use of field blanks and sample replication?

8. Have all required sampling equipment and supplies been assembled and checked?

9. Have all safety procedures been reviewed?

10. Is the timing of sample collection standardized and appropriate to the research questions? For example, is sampling focused on a summer or fall index period for lakes? Is sampling linked to seasonal climatic shifts? How many samples are collected from each site each year, and how are they distributed in time?

Some sampling activities may require obtaining special permission, particularly those activities in designated wilderness areas. It may not be possible to obtain permission to install equipment such as a stream gauge in the most highly protected areas. Plan for early coordination with land managers and natural resource professionals regarding all monitoring programs.

It is important to obtain any necessary permits or access permissions prior to finalizing the sample site list. It can also be helpful to preselect

**TABLE 2.6    KEY ISSUES TO CONSIDER IN INITIATING AND IMPLEMENTING A LONG-TERM MONITORING EFFORT TO DOCUMENT AND QUANTIFY TRENDS IN LAKE OR STREAM CHEMISTRY OVER TIME IN RESPONSE TO INPUTS OF ATMOSPHERIC DEPOSITION**

1. Include all issues listed in Table 2.5, plus the issues in this table.

2. Has a statistician or person knowledgeable about statistics been consulted in advance of carrying out the monitoring effort?

3. Has temporal variability in the subject lake or stream been characterized prior to including that body of water in the monitoring program? This might include, for example, collection of weekly or seasonal samples within the index period during 1 or 2 years.

4. Has an analysis been conducted to determine, given the amount of temporal variability documented in this water body, how large a change over what period of time would allow unambiguous, statistically significant demonstration of change over time in key water chemistry parameters?


**TABLE 2.7    KEY ISSUES TO CONSIDER IN CONDUCTING MODELING USING THE MAGIC MODEL TO ESTIMATE CRITICAL LOAD OR TO CALCULATE CHANGES IN LAKE/STREAM CHEMISTRY IN RESPONSE TO FUTURE EMISSIONS CONTROLS**

1. Include all issues listed in Table 2.5, plus those given in this table.

2. Are soils data available for the subject watershed? Although soils protocols are beyond the scope of this document, in general MAGIC requires soil chemistry data from the upper mineral B soil horizon (often the top 10 cm of the B horizon). Two to three soil pits per watershed are generally recommended. Soil parameters needed for MAGIC include pH; cation exchange capacity (CEC); exchangeable Ca, Mg, K, Na, and Al; exchangeable acidity; bulk density; loss on ignition; and an estimate of soil depth.

3. Is annual discharge available from a stream gauge at or near the location of sample collection? If not, can discharge or runoff be estimated from regional data?

4. Are estimates available for total (wet plus dry plus occult) deposition of S, oxidized N, and reduced N?

5. If models other than MAGIC are to be employed, have the required inputs been determined and are appropriate input data available?

6. If critical or target loads are to be modeled, have decisions been made regarding the resources to be protected, chemical indicators of biological effects, critical levels of those chemical indicators, and time period at which protection is to be evaluated?

backup sample locations, to be used only if access is not available to the intended sampling sites at the time of sample collection. Such an approach may be useful if, for example, road or trail access is blocked by late snowpack, road washout, avalanche, landslide, or other impediment to access of remote sites.

## 2.3.1.2 Laboratory and Sample Bottle Arrangements

Appropriate agreements will need to be made or contracts established with a qualified water chemistry laboratory well in advance of field sampling. If waters are expected to be dilute, the laboratory must be able to implement low-ionic-strength methods to achieve the necessary DQOs. The laboratory, or some other entity, should prepare and provide sample bottles, insulated shipping containers, and refrigerant.

We recommend that plasticware and plastic aliquot bottles should be high-density polyethylene (HDPE), low-density polyethylene (LDPE), or polypropylene. The sample bottle must be made of a material that is nonreactive with the chemical constituents to be measured. Polyethylene and polypropylene are commonly assumed to be essentially inert with respect to most dissolved substances. Harder plastics such as polycarbonate tend to be less reactive but will crack more easily than softer plastics. For measurement of low concentrations of dissolved carbon, glass is generally preferred. Teflon, not only the most inert plastic but also the most expensive, can be used for measurements of trace concentrations of highly reactive substances.

New bottles should be soaked in deionized water (DIW) prior to use. Samples can also be collected into previously used bottles that have been rinsed with a dilute wash acid (e.g., HCl 2%) and soaked in DIW for at least 24 h. The laboratory should follow a procedure to check acid-washed bottles to ensure that all traces of the acid are undetectable in a chemical analysis. In general, we do not recommend that bottles be acid washed. Rather, we suggest using new bottles that have been washed and subsequently checked with DIW blanks to ensure that cleaning has been adequate. Bottle processing should involve multiple rinses of both the bottle and the lid with distilled or DIW. Generally, the laboratory is responsible for providing contamination-free bottles for the sampling. Sample bottle preparation should involve triple rinsing of each bottle with DIW. The bottles should then be stored overnight, or longer, filled with DIW, followed by another rinse with DIW. Ideally, each bottle should then be filled with DIW (which can be poured out after 24 h or at the time of sample collection). Treating the sample bottles in this manner will help ensure a contamination-free sample. Laboratory conductivity analyses of blank samples typically employ a standard acceptance criterion of less than about 1.2 to 2 $\mu$S/cm.

The size of sample collection bottles can vary depending on the parameters to be analyzed but should normally be 500- or 1000-ml (large enough to allow reanalysis, if necessary), wide-mouth HDPE or LDPE bottles. For some studies, it may be possible to use smaller bottles; some studies collect 2 L of water, but this weighs about 4 pounds (1.8 kg). This can be an important constraint on sampling of multiple remote sites.

Note that if water samples are to be collected for analysis of Hg concentrations, sample bottles will need to be Teflon or glass (with Teflon-lined caps), and special bottle cleaning procedures will need to be followed, including prolonged heating in an acid solution. Check with the analytical laboratory for specific requirements.

Preprinted labels with prompts for all required information associated with the sample should be affixed to each sample bottle as part of the bottle preparation. At a minimum, this information should include (1) collection date and time, (2) site identification (lake or stream) (inlet/outlet/deep), (3) name of the person who collected the sample (first initial and complete last name), and (4) sample ID/bar code. The field crew must take precautions to ensure that no bottle mouth is contaminated with leaking refrigerant, tap water, dirt, handling contact, or other foreign substance. One reasonable precaution is to package each processed bottle individually in a zipper-lock-type plastic bag. Refrigerant should be in double zipper-lock bags. This precaution is especially important if the laboratory analysis indicates that sample contamination has been an issue in the past.

The following steps need to be completed before going to the field:

- Obtain the necessary sample bottles and (if required) syringes for each site to be sampled. Depending on the intended laboratory analyses and sample replication requirements, this can range from one to several bottles and syringes per site. Preprocessed bottles, often filled with DIW, with a preprinted or blank label tape affixed, should be provided by the analytical laboratory. It is a good idea to carry a few extra bottles beyond those needed for the intended sampling.
- Field studies often, but not always, include some sample replication (often 5% to 10% of samples) for QA purposes. Replication is generally desired to assess the repeatability of the sampling procedure, sample holding and treatment, and laboratory analysis. The amount of replication will be dependent on the sampling design and the QA/QC program. Some studies replicate all samples. This should be done in areas where access to the site is difficult and travel to the site is the biggest expense involved in the sample collection and analysis.
- Obtain ice blocks or frozen refrigerant. Ice blocks generally work better for shipping samples to the lab unless a large number of refreezable packs are used. If using refrigerant, be sure that it has been in the freezer at least 2 days prior to the day of sampling. If using block ice, it

must be placed in two securely sealed plastic bags to prevent leakage. The outer bag should be clearly marked "ice." If using refrigerant, place each refrigerant container in two securely sealed zipper-type plastic bags to prevent sample bottle contamination in the event of leakage. Place the refrigerant containers into insulated containers that will be used for sample holding and transport. Provide enough refrigerant to keep the samples cold until delivery to the lab or until placement in a refrigerator if samples are to be stored at a staging area before shipping.

- Transport the sample bottles and syringes, including the replicate bottles and process blank bottles (if applicable), in the cooler that will be used to store the samples. Bottles can be transferred to a small cooler, suitable for carrying in a backpack, prior to departing from the trailhead to access a site.

If desired, obtain one or more process blank bottles for transport to the field, followed by return with samples to the lab. We consider this step to be optional; many studies do not employ field blanks. Because the field blank bottles are generally *not* opened in the field, laboratory blanks can be an appropriate substitute that avoids the need for carrying more weight into the field.*

### 2.3.1.3  Acquisition of Equipment, Supplies, and Data Forms

Each person or sample collection team should typically be provided the following equipment and materials:

- site information folders (including maps and Stream- or Lake-Water Field-Data Forms)
- site documentation materials
- sampling protocol
- SOP documents for sampling and sample handling
- sampling bottles, and syringes if applicable
  - each bottle preferably with label affixed, placed in a zipper-lock plastic bag
  - syringes placed in a lightweight plastic box with snap-on lid, large enough to hold multiple syringes with plunger pulled three-fourths of the way out
- plastic gloves stored in a secure plastic bag
- insulated containers, refrigerant, and backpacks

---

* Note that if sample filtering is to be performed in the field (*not* recommended in this protocol), then field blanks should be transported to the field, filtered in the field, and returned to the laboratory for analysis.

- thermometer appropriate for use in air or water
- wristwatch
- survey-grade GPS and compass
- 50- or 100-m tape (and, if available, laser range finder [optional]) to measure distances
- labels and waterproof markers
- number 2 pencils, or write-in-rain-type pens, and notebooks
- digital camera with extra memory cards and battery
- heavy-duty aluminum tags, aluminum nails, and a hammer if the sites are being established for LTM of water chemistry and if this type of marking is permitted
- backpack with waterproof cover (if site is not accessible by vehicle)
- Van Dorn or other appropriate sampler if sampling a lake
- cable and instrumentation for lake "at-depth" measurements
- raft or float tube for in-lake sampling
- first aid kit
- locally determined safety equipment

Depending on the study, other materials may also be required.

Sufficient time should be allocated in advance of fieldwork for the assembly and checking of all equipment needed for the sampling program and to make sure that field personnel are thoroughly familiar with all pieces of field equipment. Arrangements will need to be made in advance for a vehicle that is suitable for the carrying capacity needs (people and equipment) and anticipated road conditions.

Supplies need to be assembled for sample collection and transport. These include sample bottles (usually provided by the laboratory), sample syringes (if applicable), refrigerant, coolers or other sample containers for transport of samples from field to vehicle, coolers for transport of samples from vehicle to field staging area, and packaging materials (including refrigerant) for shipping of samples to the laboratory.

### 2.3.1.4  Plan for Staffing

Field-sampling staffing needs should be determined well in advance of sampling activities, allowing an adequate time buffer for possible extension of the sampling effort in the event of inclement weather or unforeseen circumstances. Field efforts frequently require more time than is initially estimated. In addition, it can be advantageous to identify at least one backup field person in the event of sickness or injury.

Field personnel should be current on first aid training and local emergency procedures before heading into the field. A lead time of several months may be required to obtain proper first aid training.

## 2.3.2  Sample Collection

Sample bottles should be labeled using label tape and indelible ink. Information on the bottle label should also be recorded on a multipart chain-of-custody record (provided in Appendix D), along with information about the desired analyses and the identity of the sample collector. A field logbook should be kept in which station identification codes, date and time of sampling, and all field data are recorded. Notes on any unusual conditions at the sample sites or any circumstances that may have caused deviation from normal procedures should be recorded on the Lake or Stream Water-Sampling Record (provided in Appendix D) and described in the field data logbook.

Each sample should be labeled uniquely with site identifier, date, and sample ID/bar code in the field. An additional lab number is typically added at a later time to each aliquot in the laboratory. The sampler completes the process of filling out the bottle label at the time of sample collection. The chain-of-custody record form should include sample name, date, time of day, tests requested, comments, and appropriate signatures. Collection time should include whether the time recorded was daylight savings time or standard time. Most electronic data recording, including stream stage, is recorded in standard time, year-round. Therefore, this information is needed to relate the water sample to the electronic data.

Water sample aliquots should be collected in the field in sealed syringes or glass bottles with septum caps for some analytes to minimize contact with the atmosphere in the event that the dissolved $CO_2$ partial pressure is considerably higher than that of the atmosphere. This is a common occurrence in surface waters. Collection into a bottle having a septum cap is done by immersing and capping the bottle under water. Syringes or bottles with septum caps are used because the concentration of some analytes can change if the water sample equilibrates with atmospheric carbon dioxide partial pressure prior to analysis. DIC concentrations and pH are the measurements that are typically analyzed without contact with the atmosphere. Monomeric Al measurements are also sometimes made in this way, although the effects of $CO_2$ degassing on Al measurements are expected to be small. We therefore do not recommend collection of sample aliquots for Al analysis in syringes or septum-capped bottles.

We do, however, recommend that aliquots of samples be collected in syringes or glass bottles with septum caps along with the standard bottles, and that the syringe or septum-capped samples be used for analysis of pH and DIC. This precaution is considered to be more important for streams than it is for epilimnetic samples from lakes. If samples for these analyses are collected into bottles (with or without septum caps) in the field, it is especially important that no headspace be left in the bottle (fill completely to top) and that laboratory procedures limit the opportunity for $CO_2$ degassing in the laboratory prior to and during analysis of these parameters. Filled syringes should be transported from the field to the lab in plastic containers that minimize disturbance of the seal of the syringe.

Throughout the water chemistry sampling process it is important to take precautions to avoid contaminating the sample. Many surface waters in regions of the United States considered sensitive to effects of atmospheric deposition have low ionic strength (i.e., low levels of chemical constituents). Samples from such waters can be contaminated easily by perspiration from hands, sneezing, smoking, suntan lotion, insect repellent, fumes from gasoline engines, or chemicals used during sample collection.

For QA, sample collection should be routinely replicated so that the variability introduced by the collection process can be quantified. Although duplicate collection of samples from a subset of the sampling sites is sometimes done, the collection of three replicate samples from a subset of prespecified sites is an alternative approach for characterizing variability. The entire collection process should be repeated for the duplicate pairs or triplicates so that either two or three sample bottles representing the same sample location and approximately the same sample time are returned to the laboratory. The frequency of replicate sampling is dependent on the overall structure and requirements of the QA program. Some studies replicate all samples in the field, but they have only a subset of the replicates analyzed in the laboratory. Some studies do not include collection of field replicates. We recommend collection and laboratory analysis of duplicate pairs for 5% to 10% of the field samples in a given study. These can be used to ascertain the collective level of error or uncertainty introduced by field sampling, sample handling, short-term (seconds to minutes) temporal variation, short-distance (centimeters) spatial variation, and laboratory analyses. Thus, replicated samples provide an indication of the overall confidence that the laboratory results reflect the actual conditions in the water body at the time of sample collection.

Water temperature should be measured approximately at the location of sample collection. This can be accomplished by placing the thermometer in the water at the sampling point and waiting for the reading to stabilize. If this is not practical, temperature can be measured in a sampling bottle designated for this purpose and labeled as such. In the latter case, cool the bottle to ambient stream temperature prior to filling it with stream or lake water for immediate temperature measurement.

### 2.3.2.1   Collection of Stream Water

The collected stream water sample should be representative of stream water at the location of interest with respect to the measurements of interest. Collection of water at a single point will provide a representative sample of the channel cross section if the stream is uniformly mixed. Mixing of stream water increases with increasing flow velocity and roughness of the channel bottom.

Streams are generally well mixed with regard to dissolved substances if flow is turbulent and there are no close upstream tributaries or nearby point sources of contamination. To verify that the stream is uniformly mixed, sampling can be done for measurement of specific conductance (e.g., using a meter in the field) and perhaps other parameters at several points along the cross section and at different depths. If the measurements do not vary beyond the expected analytical variability, sampling at a single point can be done thereafter. If the required sampling location is not well mixed along the cross section, depth-integrated samples could be collected at multiple points along the cross section or the sampling site might be moved to a different location. In general, stream sampling to determine the effects of atmospheric deposition involves sampling of relatively fast-flowing small streams free of point source impacts. The water in such streams should generally be well mixed. Therefore, we recommend sampling at a single point in the main area of flow across the stream cross section unless local conditions suggest the likelihood of incomplete mixing of water in the stream.

### 2.3.2.1.1  Manual Sampling
At many site locations, the sample bottle may serve as the collection device by simply dipping the sample bottle into the stream by hand. This avoids the need for a collecting device, thereby reducing equipment needs and the chance for unnecessary sample contamination. At stream sites that are hazardous to access because of steep banks or high flows, a sampling pole (long pole that holds a bottle on the end) or a weighted bottle holder can be used so that the collection bottle can be extended out to the stream or lowered into the stream from above.

With the weighted bottle approach, an open sample bottle is placed within the weighted holder and lowered into the water with a handline. Discrete-volume samplers (such as a Van Dorn sampler) can also be used to collect the water sample but are usually not necessary in relatively small streams having well-mixed flows when the objective is measurement of dissolved constituents in stream water.

There are many acceptable methods of collecting water samples. Some, such as flow integrated stream samples, are complex and beyond the scope of the approaches described here; studies needing such sampling should consult appropriate references (e.g., USGS 2006). Methods likely to be used to collect samples for most stream studies of atmospheric deposition effects are described next. The information in this section is taken largely from the recommendations of Turk (2001).

#### 2.3.2.1.1.1  Grab/Hand Samples
Once the sampling site has been selected, bottles should be assembled and necessary information added to the labels to unambiguously identify the sample. If the bottles contain DIW, this should be discarded away from the shore so it does not disturb the sampling site.

Avoiding disturbance that can affect the water being sampled is especially important. For grab samples, the most likely disturbances are stirring up sediment or incorporating surface debris into the sample; each can contribute significant amounts of chemicals to the analytical results. Falling into the stream not only is a major disturbance of sediment but also can pose safety problems; thus, selection of a stable place to wade or a shore location from which to reach the sample location is critical. Otherwise, suitable sampling sites often are slippery because of water, ice, algae, or mud, or they contain unstable substrate such as loose boulders or poorly supported logs. If the sampler tries to use both hands for handling bottles while leaning over the water, sudden loss of balance can occur. The sample should not be collected where the sampler has waded or fallen. If the sampler is holding the bottles in hand, powder-free gloves can minimize contamination from sweat and the like. Laboratory gloves generally cover to the wrist, but longer gauntlet-style gloves cover to the elbow and should be used if the sample is collected by hand at depth greater than several centimeters. In addition to salts in sweat, common contaminants are sunscreen and insect repellent. All of these potential contaminants can be minimized by rinsing the hands and arms before collection, and rinsing the gloves, at a site far enough away (and downgradient) such that the sampling site itself is not contaminated by the rinsing.

Bottles are individually uncapped, partially filled with stream water, capped and shaken, and the rinse water discarded away from where the samples are to be collected (e.g., onshore, downstream, or where the sampler has waded). Rinse water should be poured over the cap as it is being discarded. Three rinses for each bottle are needed unless protocols otherwise indicate (e.g., bottles for total organic carbon [TOC] samples might not be rinsed). The bottles then are individually filled completely and capped. The bottles are capped underwater if septum caps are used for aliquots intended for pH and DIC analysis.

At sites where surface films contain significant pollen, insect casings, or organic film, the bottle should be kept capped until it is submerged below the surface and then uncapped to fill. The cap should then be replaced before raising the bottle back through the surface film.

The sampling depth should be consistent and documented. In general, for streams that are less than about 2-m deep, it is recommended, where possible, to sample at a depth of about 0.3 m or midway between the water surface and the water/sediment interface, whichever is closest to the surface. It often is impractical or unsafe to collect samples deeper than about 0.3 m without a sampling device of some kind.

### 2.3.2.1.2 Shallow Samples

If the water at the site is very shallow, which may be the case for many small streams and the outflow of some lakes, it may not be possible to sample very much below the surface. Very shallow streams and seeps may require creative approaches to collecting samples without disturbing sediment. It may be necessary to create a small dam that allows water to drop into the bottle. The bottle cap, pipettes, syringes, or even plastic basters used for cooking can be cleaned and used to transfer samples from the stream to the bottle in extreme cases. One option, using a syringe (use a new syringe at each site), is as follows:

1. Rinse the syringe three times with stream water, downstream of sample site as usual.
2. Use the syringe to put stream water in the sample bottle and rinse the sample bottle three times.
3. Finally, use the syringe to fill the bottle to the brim with stream water at the sample site. Cap the bottle and proceed as usual.

### 2.3.2.1.3 Pole Samples

An alternative to collecting grab samples by hand is the use of a bottle attached to a pole made of noncontaminating material such as smooth fiberglass or painted aluminum. This approach is safer than leaning over the water surface or wading and often allows the sample to be collected farther from shore and at greater depth than can be done by hand. This approach is not suitable for streams with significant velocity because of excessive drag from the assembly. The collection bottle can be larger than the sample bottle and can therefore contain sufficient water to rinse and fill it. Alternatively, the sample bottle itself can be directly attached to the pole. Because of buoyancy and leverage, it may be impractical to use a bottle larger than about 500-ml capacity and a pole longer than about 3 to 4 m. The bottle can be attached to the bottom of the pole with stainless steel hose clamps or laboratory three-finger-style bottle clamps. To minimize the possibility of contamination with surface debris or floating slush, and to allow collection at a specific depth, the bottle can be plugged with a noncontaminating silicone stopper attached to a line that the sampler pulls when the bottle is at the proper depth. The depth can be estimated or can be measured with a simple float and line attached to the pole near the bottle. Care must be taken to avoid introducing into the sample any soil or other debris that may have accumulated on the pole; it should be rinsed in an area away from the sampling site prior to use.

### 2.3.2.1.4 Deep Samples

If the stream is deeper than about 2 m, it is recommended to sample at about 0.5 m below the surface if logistics allow or to select a shallower sampling site a short distance further up- or downstream. Sample collection at a depth of 0.5 m can sometimes be achieved using a pole sampler (described in the preceding section) or a Van Dorn sampler. The choice will depend on site location and sample collection logistics. Use of a Van Dorn sampler requires low-to-moderate stream velocity and a stable position from which to collect the sample, such as a bridge, raft, or float tube.

The water sample should be collected from a point where flow velocity is high relative to other points along the cross section at the sampling location and water depth is sufficient to submerse a collection device without disturbing bottom sediments. Side pools with low velocity or eddies should not be used for sampling. Disposable, powderless latex, polyethylene, or nitrile gloves should be worn while handling sampling equipment and collecting the sample to reduce the chance of contamination. Care must be taken to avoid touching potential contaminating surfaces while wearing the gloves. Field personnel should be alert to the possibility that some individuals are allergic to contact with these glove materials.

A stream water sample can be collected from a deep stream at a specific point along the channel cross section by (1) lowering a weighted collection bottle with a handline, (2) collection with a discrete-volume water sampler, or (3) drawing the water sample with a suction pump through a tube that is lowered into the water. The sample bottle should be rinsed with stream water three times by partially filling the bottle, capping, shaking, and dumping. If wading is required, and if it is both practical and safe at that location, the sampler must stand downstream of the point of collection and avoid collecting particulates resuspended by wading or bumping the streambed with the collector. Sample collection with a tube and peristaltic suction pump can be useful when large sample volumes are needed.

### 2.3.2.1.5  Autosampling

If high-flow events need to be sampled in a nonwilderness setting, in most cases autosamplers should be used. Installation of autosampling equipment is generally not allowed in a designated wilderness. If autosampling is done during the winter at a location that experiences below-freezing temperatures, the autosampler should be kept in a heated shelter to prevent collected samples from freezing. In addition, the sampling tube should be buried between the autosampler house and the stream to prevent formation of ice plugs in the tube.

Autosampling can be done with one of several types of commercially available autosamplers. All operate similarly and consist of a controller, a peristaltic pump, sample tubing that extends into the stream, and space-efficient, custom-shaped sample bottles. The water sample is drawn through a weighted suction head that is attached to the end of the tubing. The autosampler can be set to collect at selected time intervals. This can reduce the frequency of site visits. However, to collect samples timed to the hydrograph (plot of changes in stream stage over time), the autosampler must be controlled by a programmable data logger that monitors water-level changes. Water level is most often measured by a pressure transducer that is installed in a deep portion of the stream so that it will remain below water during low flows. A large number of pressure transducers that vary in design and price are commercially available. The water level measured by the pressure transducer is typically recorded by the data logger at 15-min intervals. The data logger transfers this information to a data storage module for retrieval during site visits. The data logger can be programmed to trigger the autosampler to collect a sample based on the rate and direction of change of the water level, so that samples can be collected on ascending and descending limbs of the hydrograph as well as at the peak. Programming the data logger for this type of sampling will require some knowledge of flow variability of the stream being monitored. A weatherproof box is required to protect the data logger, storage module, and battery needed to run the data logger and autosampler. Autosamplers are usually kept in a shelter to limit the chance of vandalism.

For automatic collection of samples, the autosampler should be placed on the bank where there is no risk of it being washed into the stream during high flows. This also allows the sample tube to drain freely after sample collection. Autosampling requires that the sampling tube extend into the stream and that the suction head at the end of the tube is anchored in a deep, well-mixed portion of the channel where it will not be easily dislodged by high flows. The suction head should be positioned so that it will not draw in sediment from the bottom. The entire section of the sampling tube underwater also needs to be well anchored. The pressure transducer and its line to the data logger should be similarly anchored on the stream bottom.

Flow-activated autosamplers should be visited promptly after hydrological events to retrieve the samples and reset the autosampler. Disposable, powderless latex or nitrile gloves should be worn while handling sampling equipment to reduce the chance of contamination. Care must be taken to avoid touching potential contaminating surfaces while wearing the gloves.

On arrival at the site, bottles in the autosampler that have collected samples should be capped and labeled with the date, site, and their position number in the autosampler. These bottles are then removed and replaced with clean, empty bottles with caps removed. The autosampler is then prepared for sampling by resetting the counter and sampler spout to sample position number 1. Data are then downloaded from the data logger. These data provide the date, time, and water level associated with each sample that was collected. This information can then be used to select samples for chemical analysis.

## 2.3.2.2  Collection of Lake Water

Lake sampling should normally be done by boat, raft, or float tube over the deepest part of the lake. Such a sample is intended to represent the average lake chemistry. If boat sampling is not possible, an alternative protocol is to sample the outlet stream, if one is present, close to the lake. Outlet stream chemistry should closely approximate average lake chemistry unless there are major perturbations in the vicinity of the outlet stream. Note that even though this sample is actually collected from the outlet *stream*, it is intended to represent the chemistry of the *lake*. Thus, the sample should be labeled and documented as a lake sample rather than a stream sample. In general, we do not recommend collection of lake samples from the shoreline for the purpose of characterizing overall lake chemistry as the chemistry of this water may be different from the outlet or lake average, partly because of differences in temperature and biological productivity. Nevertheless, if there is not an outlet present, or if the outlet is not flowing at the time of sample collection, an alternative, less-desirable approach is to collect the sample from the shoreline. For this approach, a shoreline sample collection location should be selected that satisfies as many of the following criteria as possible:

- as close to the outlet as possible
- near the lowest point of land around the perimeter of the lake
- from a bedrock outcropping or otherwise-rocky area
- from the deepest accessible point

Water must be deep enough so that surface scum and sediments are not collected into the bottle, preferably in a wind-exposed area so that the water is relatively well mixed. Avoid sampling in locations having emergent vegetation or downed logs or other woody debris.

Lake samples collected from the deepest lake location are normally collected using a Van Dorn sampler at a prespecified depth. The Van Dorn sampler (and any associated tubing) should be rinsed three times with lake water and then lowered to the specified sampling depth for sample collection. The sampler should be held at the sampling depth for approximately 1 min to allow equilibration with the water at that depth. A weighted messenger is used to trigger closure of the sampler doors prior to retrieval of the sample. Water from the Van Dorn sampler is then used to rinse the sample bottle and lid (and syringes if applicable) three times prior to filling.

Sample depth for lakes should be standardized, to the extent possible, across lakes included within a particular program. For lakes deeper than 2 m, a sample depth of 1.0 or 1.5 m is commonly specified. For lakes less than 2 m deep and for lakeshore sampling (where required), the single-depth sample should be collected at 0.5-m depth. An alternate approach (not recommended as necessary for the purposes described here) is to collect a depth-integrated sample using a 2-m long tube. The EPA's NLA in 2007 collected depth-integrated samples of the euphotic zone, estimated as two times the Secchi depth, to a maximum of 2 m.

Normally, a surface sample (i.e., 1.5-m depth) from the deepest part of the lake, with or without replicates, is used to characterize lake chemistry at the time of sampling. For some studies, additional samples may be required. These might include samples of particular portions of the lake, littoral zone samples, or samples at different depths. Even if samples are not collected at different depths, lake sampling should ideally be accompanied by measurements of the temperature profile to determine if the lake is stratified and to characterize the location and depth of the thermocline, the epilimnion, and the hypolimnion. If samples are collected at different depths, the water temperature should be measured at each sampled depth in conjunction with measuring the temperature profile. Water samples cannot be determined to be epilimnetic samples unless a depth profile is taken. Surface samples without temperature profile should be labeled as water surface samples rather than epilimnetic samples to avoid possible confusion.

Point samples are those collected at a specific depth. Van Dorn cylinders and Kemmerer bottles are the most common point samplers for lakes. The Van Dorn cylinder may have some advantage in that it appears to allow better circulation of water through the sample container. In general, we recommend use of a Van Dorn sampler for lake sampling. In either case, the sampler is difficult to keep clean unless it is kept in a plastic bag between sites. The sampler should be soaked in the lake prior to use. At the sample site, the sampler should be raised and lowered several times just below the surface to further rinse the container. It is then lowered to the desired depth, held to stabilize, and triggered, usually with a weight that slides down the line holding the sampler. When sampling the hypolimnion, care should be taken not to touch the bottom because this will disturb sediment that could contaminate the sample.

Because of the drag of long lengths of rope and the sampler itself, both Van Dorn and Kemmerer samplers are prone to sampling at shallower depths than indicated by the length of the rope. If the boat is drifting because of current or wind, deeper samples may be in error by a considerable margin for reported depth. This error can be avoided by anchoring or tying to a buoy. These samplers also tend to plane while being lowered; allowing the rope to straighten before triggering the bottle can help minimize this error.

Pumps and tubing sometimes are used to collect point samples from lakes and streams or to integrate samples from lakes. The primary concern with these devices is keeping the tubing clean. It is impossible to thoroughly clean the inside of a tube to eliminate bacterial growth. Sampling tubes can generally be kept clean by storing them filled with DIW and in a black plastic bag and avoiding disturbance of sediment while sampling. The tubing can easily block with slush during freezing conditions. Tubing does offer the ability to collect an integrated sample of the water column of lakes by lowering the tubing, at a constant rate of travel while pumping, to near the bottom and back to the surface. In general, however, lake water sampling for inventory and monitoring of the effects of atmospheric deposition does not require collection of depth-integrated samples with a tube.

## 2.3.3   On-Site Measurements

### 2.3.3.1   Evaluation of Site Characteristics

The field notebook and all field forms should be filled in while personnel are at the sampling location so the sample can be accurately linked to field data and observations. A sample field form is shown in Appendix D. Field notebooks are helpful at sites used for LTM to provide easy access to locational information and maps, historical information on site characteristics, and field data collected during previous years.

Observations and impressions made by the field teams at the sampling location and elsewhere on the target stream or lake are extremely useful for ecological value assessment, evaluation of general water body condition, and data verification and validation. Thus, it is important that observations made by the field team about lake, stream, or watershed characteristics and condition be recorded while the field personnel are in the field. Field data forms are available, and field notebooks should be provided for this purpose. The forms are designed as a guide for recording pertinent field observations. Field data entry forms are never considered to be comprehensive. Any additional observations made by the field crew that might eventually be useful in making a site condition assessment should be recorded in the "Comments" section of the field notebook. Team members complete the form at the end of the sampling, taking into account all observations made while on site.

### 2.3.3.2   Stream Stage and Discharge and Lake Level

#### 2.3.3.2.1   Stream Stage and Discharge

The most valuable nonchemical measurement for interpreting stream chemistry data is often stream flow, which is measured as volume of water per unit time (also referred to as discharge). Variations in stream flow reflect precipitation and the different pathways water takes to reach the stream channel. During low-flow conditions, water discharging into the stream channel has usually had opportunity to pass well below the surface soil into deeper soils, till, or bedrock. Such deeper flow paths provide greater contact between water and the soils and geologic materials in the watershed. As a consequence, base flow commonly receives larger quantities of weathering products that can buffer acidity, raise pH, and increase concentrations of BC in solution. During high-flow periods, some water enters the stream channel through shallow flow paths that more clearly reflect the chemistry of upper soil horizons. Shallow flow paths tend to result in lower concentrations of BC in drainage water, less acid neutralization (sometimes increased acidity), and higher concentrations of DOC than deeper flow paths.

The USGS is the recognized leader in development and implementation of flow measurements. USGS protocols and recommended equipment for measuring flow are detailed by Rantz et al. (1982). This section provides a synopsis of this material. In addition, the USGS has produced a training video for measuring discharge (http://training.usgs.gov/TEL/Nolan/SWProcedures/Index.html). Field personnel may want to review this video as part of their field training program.

To measure stream flow, some type of channel control is necessary. This control may be constructed as a temporary feature, such as a weir or dam, or a natural control, such as a bedrock outcrop or channel-width restriction. An effective control provides a predictable relationship between water level (stage) and flow (discharge) that does not change over time. A pressure transducer installed in the deepest part of the stream channel, just upstream of the channel control, can be used to record the water level, commonly at 15-min intervals. A line must be secured in the stream to transmit the response of the pressure transducer to a data logger. Thus, the pressure transducer measures changes in stage; these stage measurements then must be converted to estimates of flow. Alternatively, stage can be measured using a measuring rod or yardstick held vertically in place at a specific location. That location must be clearly defined using a permanent structure of some sort, such as a rock or large tree.

To establish the relationship between stage and flow (referred to as a rating curve), simultaneous stream stage and flow measurements are needed over as wide a range of stream flows as possible. To conduct the stream flow measurements, a cross section is chosen in the general vicinity of where the water-level measurements are taken. The ideal cross section chosen for measurement should provide a regular cross-sectional channel shape that provides laminar flow throughout the channel. The more closely these conditions are met, the more accurate will be the resulting estimates of discharge.

The cross section is divided into intervals such that at least one pair of depth and stream velocity measurements can be made in each interval. The number and width of the intervals are dependent on the shape of the cross section. Stream flow measurement determined with a single stream velocity measurement is not sufficient for obtaining an accurate representation of discharge.

An additional common method of estimating stream velocity relies on measuring the velocity of a neutrally buoyant object, such as a small orange, traveling downstream. This approach can provide grossly inaccurate flow estimates. The object can follow preferentially rapid flow paths or, conversely, be temporarily impeded by stones or wood in the channel. We do not recommend use of this method for estimating stream discharge, although it can be useful for instructional purposes.

The stream velocity can vary considerably along the cross section and with depth, requiring a number of velocity measurements to obtain an accurate flow measurement. A large variety of stream velocity meters with varying precision and accuracy are available commercially. The velocity of each interval is multiplied by the cross-sectional area of that interval, and the products of all intervals are summed to provide the overall estimate of stream flow.

The salt dilution gauging method also provides an approach for estimating stream flow at locations where reliable measurements with flow meters are not possible. Such conditions occur in streams that are highly turbulent or have irregular channels, or in small streams or during low flow under conditions when a large part of the flow is through gravel and rocks in the stream bed. For such conditions, salt dilution gauging provides a more reliable method for discharge measurement. Salt dilution gauging involves the addition of a known quantity of salt upstream of the gauging site, either by a single addition or by continuous injection. Discharge is computed based on the concentration or dilution of the salt, determined by conductivity measurements, as it passes the gauging site. We do not recommend routine use of this method, and it may not be allowed by land owners or managers.

Making flow measurements during periods of high flow may not be safe, or it may not be possible to wade into the stream under such conditions. If there is any doubt about the safety of wading under the existing flow conditions, field staff should *not* enter the stream.

If information on flow is needed to aid in the interpretation of stream chemistry measurements, but neither installation of a stream gauge nor collection of flow measurements are feasible, water level can be manually recorded from a staff gauge at the time that the water sample is collected. This will provide data on stage but not discharge. For some research or monitoring project objectives, relative differences in stream stage may be sufficient in place of the more quantitative discharge data. The staff gauge (not allowed in wilderness settings) should be located just upstream of an effective control. The staff gauge is usually a pressure-treated post or metal fence post anchored in the stream with a large ruler attached. The elevation of the ruler should be surveyed in reference to an object near the bank that would be considered immovable. This enables future verification that the staff gauge has not moved. If the chemical concentrations that are being measured are statistically related to flow, they are also related to stage, although the relationships can differ. Changes in chemical concentrations of flow-dependent constituents can be estimated from stream stage measurements in a manner similar to that done with flow measurements.

Another approach is to characterize regional flow conditions based on nearby gauges on similar watersheds. Using this approach, it is possible to obtain a general idea of the likely flow conditions at the sampling site at the time of sample collection without the need for site-specific measurements. It is important to note that, although discharge can be a useful parameter in evaluating the effects of atmospheric deposition, it is not absolutely essential to have site-specific discharge data. One should not choose to avoid sampling a site for water chemistry simply because it is impractical or impossible to collect parallel data on discharge.

### 2.3.3.2.2  *Lake Level*

For interpreting lake chemistry data, the lake level can be especially useful. Of particular importance is the likelihood that lake chemistry will vary with precipitation cycles. During drought periods, a higher proportion of inflowing water may follow relatively deep flow paths, allowing for greater acid neutralization and BC mobilization. During wet periods, drainage water may preferentially follow shallow flow paths, allowing less contact with soils and geologic materials and therefore limited acid neutralization. Seepage lakes may receive proportionately greater inflow of groundwater (which, depending on geological and soil conditions, may be rich in BC) during drought periods. It is also possible that some seepage lakes might lose their connection with the groundwater during drought, causing the opposite effect.

The extent of such influences on hydrology and consequent acid neutralization is expected to be region and watershed specific. Measurement of lake level at the time of sampling can provide critical data to help sort such effects. The simplest way to collect such data is to install a fixed staff gauge in the lake and record the lake level at each sampling time. In wilderness settings, or other locations where installation of a staff gauge is not allowed or is impractical, relative lake level can be documented by measuring the vertical height of the lake surface below a fixed landmark (e.g., a large shoreline rock or tree root).

## 2.3.3.3  Ancillary Data (Chemical and Physical Information That Is Not Necessary But May Be Useful)

### 2.3.3.3.1  *Chemical Data*

Physicochemical data can be collected on site with field equipment for measurements such as pH, specific conductance, DO, and turbidity. This information can be useful for reconnaissance work and sample site selection or for other investigations in which real-time information is needed to direct field activities. In general, however, we recommend that assessments or monitoring of acid-base chemistry or nutrient status should be made using chemical analyses conducted in the laboratory rather than the field. Commercially available equipment can produce data of quality similar to that of laboratory equipment for some variables. However, reproducing the clean, controlled environment of a laboratory in the field is difficult. Therefore, if real-time data are not required to satisfy the objectives of a particular study or other requirements,* conducting the chemical analysis in the laboratory is recommended to ensure high data quality. If chemical analysis will be required in the field, premobilization and field calibration checks should be conducted. Other types of chemical measurements can

---

\* For example, *in situ* measurement of pH is sometimes required for data used by state 303(d) water quality assessment programs to determine waters to be classified as water quality limited according to the Clean Water Act.

also be made in the field with commercially available analysis kits, but the data obtained using these methods typically only provide a rough approximation that may not be sufficiently accurate or precise for inventory or monitoring.

Meters with probes that continuously monitor pH and specific conductance are also available. These probes effectively characterize temporal variability but lack the precision and accuracy of laboratory measurements. The need for temporal resolution should be weighed against DQOs, logistics, and costs to determine if *in situ* monitoring is advantageous.

### 2.3.3.3.2  Physical Data

Interpretation of water chemistry data can be significantly aided by ancillary physical measurements such as air and water temperature, weather conditions, recent precipitation, and snowpack water equivalence (Table 2.8). Additional data might also be collected at the sampling site, depending on the study.

For example, it can be useful to develop lake thermal profiles to evaluate the extent of lake stratification. For stratified lakes, it can be useful to collect water samples from the hypolimnion. Such data can be used in evaluating S reduction in the lake sediment, hypolimnetic DO depletion, or the dynamics of P retention and release in lake sediments. Unless detailed analyses of S or P cycling are to be conducted, however, hypolimnetic samples are generally not needed.

**TABLE 2.8  ANCILLARY MEASUREMENTS THAT MAY HELP IN INTERPRETATION OF LAKE OR STREAM WATER CHEMISTRY DATA**

| Streams | Lakes |
|---|---|
| • Discharge or stage | • Secchi disk transparency |
| • Water temperature | • Thermal profile |
| • Air temperature | • Chlorophyll *a* |
| • Snowpack depth and snow water equivalence | • Level (if lake is to be sampled multiple times) |
| • Precipitation | • Hypolimnetic water samples |
| • Watershed morphometry | • Littoral zone water samples |
| • Fish stocking and management | • Dissolved oxygen |
| • Watershed disturbance[a] | • Presence of inflowing or outflowing streams |
| | • Watershed morphometry |
| | • Fish stocking and management |
| | • Watershed disturbance |

[a] Watershed disturbance can be evaluated by field reconnaissance, examination of aerial photos, regional land cover data sets, and so on.

Studies of temporal trends in surface water quality or characterization of water quality conditions within a specific lake or stream or across a park or forest can be designed to assess a variety of kinds of impacts and changes in those impacts over time. For this to be successful, information is often needed regarding basic watershed features that can have an impact on water quality. These can include, for example, the variables listed in Table 2.8.

The ecological significance of aquatic ecosystem degradation and loss caused by physical habitat alterations can, in some cases, exceed degradation caused by atmospheric deposition or human activities that affect water chemistry. Therefore, physical habitat surveys of lakeshore areas, littoral zones, stream channels, and riparian zones can be useful in conducting overall habitat condition assessments and in interpreting water chemistry data. Habitat information is helpful in the interpretation of what lake or stream biological assemblages "should" be like in the absence of many types of anthropogenic impacts. The physical evaluation can provide a reproducible, quantified estimate of habitat condition, serving as a benchmark against which to compare future habitat changes that might result from anthropogenic activities or extreme events. Furthermore, habitat information can aid in the diagnosis of probable causes of ecological impairment in lakes or streams.

In addition to information collected in the field by the shoreline, stream channel, or littoral zone surveys, the physical habitat description of each lake or stream can include many map-derived or measured variables such as lake surface area, shoreline length, stream width-to-depth ratio, and habitat integrity or complexity. Furthermore, an array of information, including watershed topography and land use, supplements the physical habitat information. The shoreline, channel, and littoral surveys concentrate on information best derived "on the ground." As such, these survey results provide part of the linkage between large watershed-scale influences and those forces that directly affect aquatic organisms day to day. Together with water chemistry, the habitat measurements and observations describe the variety of physical and chemical conditions that are necessary to support biological diversity and foster long-term ecosystem stability.

Habitat surveys should not be considered a *necessary* component of inventory and monitoring. They require a commitment of time and resources. Nevertheless, the data collected in such surveys can be helpful in the subsequent interpretation of effects, especially if the documentation or quantification of effects relies on collection of biological (i.e., phytoplankton, zooplankton, benthic invertebrates, fish) as well as chemical variables.

The shoreline and littoral habitat surveys conducted by the EPA in the EMAP program employed a randomized, systematic design with 10 equally spaced observation stations located around the shore of each sample lake. Teams went to the field with premarked lake outlines showing the locations of these stations. The observations at each station included quantitative and semiquantitative observations of vegetation structure, anthropogenic disturbances, and bank substrate. In-lake littoral measurements and onshore observations dealt with littoral water depth, bottom substrate, nearshore fish cover, and aquatic macrophyte cover. With quantifiable confidence, investigators condensed these observations into descriptions applicable to the whole lakeshore and littoral zone. For example, team observations led to quantitative descriptions such as the mean canopy or aquatic macrophyte cover along the lakeshore, the extent of shoreline disturbed by various human activities, and the dominant littoral substrate in the lake. There are similar physical habitat evaluation procedures for streams, developed for EPA's national surveys, such as the WSS.

### 2.3.3.4   On-Site Processing of Samples

In general, we do not recommend filtering lake or stream samples in the field except where immediate filtering is required for a particular measurement. To avoid the possibility of sample contamination, it is generally preferable to perform this step within the controlled conditions of a laboratory. Similarly, measurement of pH in the field is not recommended for most studies. This measurement is best performed in the laboratory under controlled conditions.

For most chemical constituents of interest for atmospheric deposition studies, sample preservation in the field is not necessary. Types of sample preservation may include addition of chemicals or filtering to remove particulates. Preservation procedures are generally done for a specific measurement and usually render the sample unusable for other measurements. Therefore, if preservation in the field is needed for a specific measurement, an aliquot will need to be removed from the sample prior to preservation. The volume of the aliquot will be dependent on the analytical requirements of the measurement. In general, we do not recommend sample preservation in the field for chemical analyses.

Most samples contain dissolved or particulate organic matter and associated microbes that can change sample chemistry through decomposition and assimilation. All samples should therefore be placed out of the sunlight and in a cooler with ice as soon as possible for transport back to the laboratory where they can be refrigerated. This procedure is usually sufficient to slow biological processes enough to prevent measurable changes in chemical concentrations. As a rule of thumb, samples should be returned from the field as quickly as possible to enable processing in the laboratory, and filtration (if needed) should be performed in the laboratory. An exception applies to the collection and analysis of samples for measurement of chlorophyll $a$. Those samples are most commonly filtered in the field, and the filter (not the filtrate water) is transported in a zipper-lock bag on ice in the dark to the laboratory for analysis. It is essential to record in the Notes section of the Water-Sampling Record Form the volume of water that was filtered for chlorophyll *a* measurement.

## 2.3.4 Postcollection Sample Processing, Documentation, and Cleanup

Filled sample bottles should be placed in zipper-lock bags prior to transport from the field site. Syringes should be packed in a generic lightweight plastic box with snap-on lid. Each box needs to be long enough to hold syringes that are two-thirds to three-quarters filled with sample water and wide enough to hold multiple syringes. The bagged sample bottles and boxed syringes should be packed with double-bagged ice or frozen refrigerant for transport.

Insulated containers, with double-bagged chemical refrigerant ("blue ice"), or preferably with double-bagged ice blocks, are needed for transport of collected samples between the field and other staging location and eventually to the laboratory. Ice works better for shipping unless large numbers of chemical refrigerant packs are used. Small insulated containers that will fit into backpacks can be used to carry and protect the samples in the field. Coolers can be used for assembly and transport of samples in vehicles.

Chemical refrigerant containers should be packaged in two zipper-lock plastic bags to minimize the possibility of sample bottle contamination through leakage. The field crews will need to make sure that the chemical refrigerant is placed in a freezer at least 2 days before sampling.

### 2.3.4.1 Sample Documentation

Sample documentation should be completed in the field. It will include completing and affixing all sample labels, completing all field and chain-of-custody forms, and recording field notes and site condition information. A list of suggested minimum database requirements for studies of the effects of atmospheric deposition are provided in Table 2.9. Documentation should also include review of sampling procedures, labeling, and photographic/written documentation (Table 2.10).

### 2.3.4.2 Postsampling Equipment Cleanup

Cleanup is important to minimize the possibility of transporting pathogens, noxious species, or invasive species from one sampling location to another. Risks vary from region to region and location to location. Field personnel should consult with regional and local land management offices for specific problem identification and appropriate precautions and cleaning protocols. A variety of forest pathogens and aquatic and terrestrial invasive species may be of concern, depending on location. Personnel and their boots, vehicles, boats, and equipment can serve as transport vehicles for problematic species.

**TABLE 2.9    LIST OF SUGGESTED MINIMUM[a] DATABASE REQUIREMENTS FOR THE SURFACE WATER CHEMISTRY MONITORING RECORD, CHAIN OF CUSTODY, AND SITE SUMMARY**

- Lake/stream ID
- Official USGS lake/stream name and alternative name/field ID
- Latitude (decimal degrees)
- Longitude (decimal degrees)
- Datum used (use North American Datum [NAD]83 if possible)
- Long-term or synoptic sampling program
- Technician responsible for field activities
- Date
- Time
- Sample type (regular, duplicate, blank, split, etc.)
- Sample measurement location (i.e., inlet, outlet, deep, shore [lake], bank [stream], etc.)
- Sample method (i.e., grab, pole, *in situ*, etc.)
- Sample ID/bar code
- Usual collection point (yes, no). Why not?
- Protocol deviation (yes, no). Why?

[a] There may be other requirements according to the individual study. Note that this protocol does not *require* collection of temperature or stream flow data, although it does recognize that such data can be useful in interpreting the results of chemical analyses.

**TABLE 2.10    KEY ISSUES TO CONSIDER AFTER COMPLETING FIELD SAMPLING**

1. Were the samples collected at the designated depth and free of influence from any sediment that could be disturbed during sample collection?
2. Were all sample bottle and sample syringe labels fully completed with all required information?
3. Was the sampling fully documented, including site photographs (if appropriate), completed field and chain-of-custody forms, and field notes?
4. Were any conditions or circumstances noted that could potentially compromise or influence the chemistry of the sample?

Proper cleaning of equipment, boots, and so on should be done before leaving the site. Wherever a risk exists, field personnel should take additional site-specific appropriate risk management precautions.

## 2.3.5 Sequence of Field Activities

The recommended sequence of field activities to be conducted by field personnel at the sampling site is as follows:

1. Select/verify sampling site location
2. Take photographs
3. Fill out and affix the label for each sampling bottle to be filled at the site if you have not already done so
4. Evaluate and document site conditions
5. Verify how many and what kinds of samples will be collected (QA replicates, single versus integrated sample, special aliquots to be collected into glass bottle or syringe)
6. Collect water samples
7. Preserve (if necessary) or transfer to glass or syringe (as appropriate) selected sample aliquots
8. Place collected samples in cold, dark storage container
9. Collect any needed ancillary data
10. Determine (if appropriate) stream discharge or stage
11. Complete all site documentation and chain-of-custody forms
12. Record all field observations in field notebook
13. Clean equipment, clothing, and boots to prevent spreading invasive species to another site

## 2.3.6 Safety in Field Activities

### 2.3.6.1 Key Safety Considerations

For safety reasons, an emergency contact individual who is not part of the field crew should always know where the field crew is going each day and by what route. This person should be contacted by the field crew immediately on return from the field each day.

For sampling remote locations, safety equipment should include, but should not necessarily be limited to, the following:

- two-way radios or cellular telephone (if cell phone access is available in the study area)
- extra batteries for GPS and radios
- rain gear

- emergency shelter blanket
- adequate supply of drinking water or appropriate water filtration system
- sunscreen
- first aid kit
- locally required safety equipment

All field personnel should have current first aid and cardiopulmonary resuscitation (CPR) certificates. Field personnel should never enter a deep or fast-flowing stream without wearing a personal flotation device (PFD). A PFD should also be worn when working close to a stream during high-flow conditions. If safety is in doubt, *do not* collect the sample.

When sampling a lake at the midlake location from a float tube or raft, it can be advantageous for one person to remain on the shore to provide logistical support, to record data, and as a safety precaution. If necessary, a two-way radio can be used to facilitate communication between midlake and shore. Canoe sampling is more commonly done with two people in the canoe. Throughout the sample collection process, the person in the boat should be careful to avoid drifting from the sampling site. The sampler can evaluate position by keeping two onshore landmarks in line. The onshore person should continuously check the float tube location.

## 2.3.6.2  Job Hazard Analysis

Field personnel should review potential job hazard information prior to going into the field and should check with land managers regarding safety concerns and existing job hazard analyses for the study area. They should also construct a field itinerary prior to fieldwork. The itinerary should include

- departure date and time
- expected return date and time
- expected route of travel (roads, trailheads, trails, destinations)

The job hazard information should address items such as the following:

| | |
|---|---|
| • disorientation while in backcountry | • field clothing |
| • personal gear | • exposure |
| • environmental hazards | • first aid |
| • drinking water | • animal encounters |
| • horse riding and management issues | • severe weather |
| • stream crossing | • emergency evacuation |
| • vehicle safety and road access blockage | • safety equipment |

# REFERENCES

Aston, L. and J.N. Seiber. 1997. The fate of summertime airborne organophosphate pesticide residues in the Sierra Nevada mountains. *J. Environ. Qual.* 26(1483–1492).

Baker, J.P. and C.L. Schofield. 1982. Aluminum toxicity to fish in acidic waters. *Water Air Soil Pollut.* 18:289–309.

Baron, J.S. 2006. Hindcasting nitrogen deposition to determine ecological critical load. *Ecol. Appl.* 16(2):433–439.

Baron, J.S., D.S. Ojima, E.A. Holland, and W.J. Parton. 1994. Analysis of nitrogen saturation potential in Rocky Mountain tundra and forest: implications for aquatic systems. *Biogeochemistry* 27:61–82.

Borum, J. 1996. Shallow waters and land/sea boundaries. In B.B. Jorgensen and K. Richardson (Eds.), *Eutrophication in Coastal Marine Ecosystems.* American Geophysical Union, Washington, DC, pp. 179–203.

Bricker, S.B., C.G. Clement, D.E. Pirhalla, S.P. Orlando, and D.G.G. Farrow. 1999. *National Estuarine Eutrophication Assessment: Effects of Nutrient Enrichment in the Nation's Estuaries.* Special Projects Office and the National Centers for Coastal Ocean Science, National Ocean Service, National Oceanic and Atmospheric Administration, Silver Spring, MD.

Burns, D.A. 1996. Retention of $NO_3^-$ in an upland stream environment: a mass balance approach. *Biogeochemistry* 40:73–96.

Campbell, D.H., D.W. Clow, G.P. Ingersoll, M.A. Mast, N.E. Spahr, and J.T. Turk. 1995. Processes controlling the chemistry of two snowmelt-dominated streams in the Rocky Mountains. Water Resour. Res. 31(11):2811–2821.

Colquhoun, J.R., W.A. Kretser, and M.H. Pfeiffer. 1984. *Acidity Status Update of Lakes and Streams in New York State.* New York State Department of Environmental Conservation, Albany.

Cook, R.B., J.W. Elwood, R.R. Turner, M.A. Bogle, P.J. Mulholland, and A.V. Palumbo. 1994. Acid-base chemistry of high-elevation streams in the Great Smoky Mountains. *Water Air Soil Pollut.* 72:331–356.

Cronan, C.S. and C.L. Schofield. 1979. Aluminum leaching response to acid precipitation: effects on high elevation watersheds in the Northeast. *Science* 204:304–306.

Dickson, W.T. 1978. Some effects of the acidification of Swedish lakes. *Verh. Int. Ver. Theor. Angew. Limnol.* 20:851–856.

Driscoll, C.T. and J.J. Bisogni. 1984. Weak acid/base systems in dilute acidified lakes and streams of the Adirondack region of New York State. In J.L. Schnoor (Ed.), *Modeling of Total Acid Precipitation.* Butterworth, Boston, pp. 53–72.

Driscoll, C.T., J.P. Baker, J.J. Bisogni, and C.L. Schofield. 1980. Effect of aluminum speciation on fish in dilute acidified waters. *Nature* 284:161–164.

Elser, J.J., T. Andersen, J.S. Baron, A.-K. Bergström, M. Jansson, M. Kyle, K.R. Nydick, L. Steger, and D.O. Hessen. 2009. Shifts in lake N:P stoichiometry and nutrient limitation driven by atmospheric nitrogen deposition. *Science* 326:835–837.

Eriksson, E. 1981. Aluminum in groundwater, possible solution equilibria. *Nord. Hydrol.* 12:43–50.

Eshleman, K.N. 1988. Predicting regional episodic acidification of surface waters using empirical techniques. *Water Resour. Res.* 24:1118–1126.

Fenn, M.E., R. Haeuber, G.S. Tonnesen, J.S. Baron, S. Grossman-Clark, D. Hope, D.A. Jaffe, S. Copeland, L. Geiser, H.M. Rueth, and J.O. Sickman. 2003. Nitrogen emissions, deposition, and monitoring in the western United States. *BioScience* 53(4):391–403.

Fenn, M.E., M.A. Poth, and D.W. Johnson. 1996. Evidence for nitrogen saturation in the San Bernardino Mountains in southern California. *For. Ecol. Manage.* 82:211–230.

Gilliam, F.S., M.B. Adams, and B.M. Yurish. 1996. Ecosystem nutrient responses to chronic nutrient inputs at Fernow Experimental Forest, West Virginia. *Can. J. For. Res.* 26:196–205.

Green, R.H. 1979. *Sampling Design and Statistical Methods for Environmental Biologists.* Wiley, New York.

Helsel, D.R. and R.M. Hirsch. 1992. *Statistical Methods in Water Resources. Studies in Environmental Science 49.* Elsevier, New York.

Kahl, J.S., J.L. Stoddard, R. Haeuber, S.G. Paulsen, R. Birnbaum, F.A. Deviney, J.R. Webb, D.R. Dewalle, W. Sharpe, C.T. Driscoll, A.T. Herlihy, J.H. Kellogg, P.S. Murdoch, K.M. Roy, K.E. Webster, and N.S. Urquhart. 2004. Have US surface waters responded to the 1990 Clean Air Act Amendments? *Environ. Sci. Technol.* 38:485A–490A.

Kopp, B.S. and H.A. Neckles. 2004. *Monitoring Protocols for the National Park Service North Atlantic Coastal Parks: Estuarine Nutrient Enrichment.* Report submitted to National Park Service. USGS Patuxent Wildlife Research Center, Augusta, ME.

Lawrence, G.B. 2002. Persistent episodic acidification of streams linked to acid rain effects on soil. *Atmos. Environ.* 36:1589–1598.

Lawrence, G.B., B. Momen, and K.M. Roy. 2004. Use of stream chemistry for monitoring acidic deposition effects in the Adirondack region of New York. *J. Environ. Qual.* 33:1002–1009.

Lawrence, G.B., K.M. Roy, B.P. Baldigo, H.A. Simonin, S.B. Capone, J.W. Sutherland, S.A. Nierzwicki-Bauer, and C.W. Boylen. 2008. Chronic and episodic acidification of Adirondack streams from acid rain in 2003–2005. *J. Environ. Qual.* 37:2264–2274.

McCune, B., J. Grenon, L.S. Mutch, and E.P. Martin. 2007. Lichens in relation to management issues in the Sierra Nevada national parks. *Pacific Northwest Fungi* 2(3):1–39.

Melack, J.M., S.C. Cooper, T.M. Jenkins, L. Barmuta, Jr., S. Hamilton, K. Kratz, J. Sickman, and C. Soiseth. 1989. *Chemical and Biological Characteristics of Emerald Lake and the Streams in Its Watershed, and the Response of the Lake and Streams to Acidic Deposition.* California Air Resources Board, Sacramento.

Muniz, I.P. and H. Levivestad. 1980. Acidification effects on freshwater fish. In D. Drabløs and A. Tollan (Eds.). *Proceedings of the International Conference on the Ecological Impact of Acid Precipitation.* SNSF Project, Oslo, Norway. pp. 84–92.

Murdoch, P.S. and J.B. Shanley. 2006. Flow-specific trends in river-water quality resulting from the effects of the Clean Air Act in three mesoscale, forested river basins in the northeastern United States through 2002. *Environ. Monitor. Assess.* 120(1–3):1–25.

Murdoch, P.S. and J.L. Stoddard. 1992. The role of nitrate in the acidification of streams in the Catskill Mountains of New York. *Water Resour. Res.* 28(10):2707–2720.

Nixon, S.W. 1995. Coastal marine eutrophication: a definition, social causes, and future concerns. *Ophelia* 41:199–219.

Rantz, S.E., and others. 1982. *Measurement and Computation of Streamflow. Measurement of Stage and Discharge.* Vol. 1. US Geological Survey Water Supply Paper 2175. US Government Printing Office, Washington, DC.

Schofield, C.L. and J.R. Trojnar. 1980. Aluminum toxicity to brook trout (*Salvelinus fontinalis*) in acidified waters. In T.Y. Toribara, M.W. Miller, and P.E. Morrows (Eds.), *Polluted Rain*. Plenum Press, New York, pp. 341–362.

Seip, H.M. 1980. Acidification of freshwater—sources and mechanisms. Ecological impact of acid precipitation. In *Proceedings of an International Conference, SNSF Project*, Sandefjord, Norway. Oslo, Norway, pp. 358–365.

Sickman, J.O., A. Leydecker, C.C.Y. Change, C. Kendall, J.M. Melack, D.M. Lucero, and J. Schimel. 2003a. Mechanisms underlying export of N from high-elevation catchments during seasonal transitions. *Biogeochemistry* 64:1–24.

Sickman, J.O., J.M. Melack, and D.W. Clow. 2003b. Evidence for nutrient enrichment of high-elevation lakes in the Sierra Nevada, California. *Limnol. Oceanogr.* 48(5):1885–1892.

Stoddard, J.L. 1994. Long-term changes in watershed retention of nitrogen: its causes and aquatic consequences. In L.A. Baker (Ed.), *Environmental Chemistry of Lakes and Reservoirs*. American Chemical Society, Washington, DC, pp. 223–284.

Sullivan, T.J., J.R. Webb, K.U. Snyder, A.T. Herlihy, and B.J. Cosby. 2007. Spatial distribution of acid-sensitive and acid-impacted streams in relation to watershed features in the southern Appalachian mountains. *Water Air Soil Pollut.* 182:57–71.

Turk, J.T. 2001. *Field Guide for Surface Water Sample and Data Collection*. Air Program, USDA Forest Service, Washington, DC.

Ulrich, B., R. Mayer, and T.K. Khanna. 1980. Chemical changes due to acid precipitation in a loess-derived soil in central Europe. *Soil Sci.* 130:193–199.

US Geological Survey. 2006, September. Collection of water samples (ver. 2.0). In *US Geological Survey Techniques of Water-Resources Investigations*, Book 9, Chapter A4. http://pubs.water.usgs.gov/twri9A4/, accessed March 10, 2014.

Valiela, I., K. Foreman, M. LaMontagne, D. Hersh, J. Costa, P. Peckol, B. DeMeo-Andreson, C. D'Avanzo, M. Babione, C. Sham, J. Brawley, and K. Lajtha. 1992. Couplings of watersheds and coastal waters: sources and consequences of nutrient enrichment in Waquoit Bay, Massachusetts. *Estuaries* 15:443–457.

Williams, M.W., J.S. Baron, N. Caine, R. Sommerfeld, and J.R. Sanford. 1996. Nitrogen saturation in the Rocky Mountains. *Environ. Sci. Technol.* 30:640–646.

Zabik, J.M. and J.N. Seiber. 1993. Atmospheric transport of organophosphate pesticides from California's Central Valley to the Sierra Nevada Mountains. *J. Environ. Qual.* 22:80–90.

# Chapter 3

# Laboratory Analyses

## 3.1 INTRODUCTION

Surface water samples collected in watersheds on forested lands with minimal recent influence from urban or agricultural development tend to have low concentrations of nutrients. Because these watersheds are most commonly located in upland areas with rocky, infertile soils, their drainage waters also tend to have low ionic strength. This means that most other dissolved constituents are also low in concentration. Analysis of water samples having low concentrations of major constituents is challenging. Chemical analysis of dissolved materials at low concentrations sometimes requires modification of the standard methods used by the laboratory for routine sample preparation and analysis. The laboratory selected for analyzing low-nutrient, low-ionic strength surface water samples should not only be experienced with these types of samples but also should (1) analyze them on a routine basis and (2) provide data with reporting limits sufficiently low for project needs.

Studies of effects of atmospherically deposited constituents, including acid precursors, nutrients, and toxic substances, entail collection of water samples and laboratory analysis of a wide range of chemical parameters using a wide range of instruments and analytical methods. It is beyond the scope of this book to describe them. Nevertheless, there are a number of laboratory issues common to most or all such investigations. These are the issues we focus on in presenting this protocol. We outline important steps to be taken, or at least considered, by the laboratory in conducting water quality studies. These include elements of laboratory preparation prior to sample analysis and the analyses themselves. In preparing to accept samples for processing, the laboratory must rely on developed procedures for bottle cleaning, sample processing, and the preservation (as needed) and storage of samples in the laboratory. The investigators must ensure that, to the extent possible, preprocessing holding times prior to sample analysis are not exceeded. The chemical analyses must focus on adherence to the quality assurance/quality control (QA/QC) plan, outlined in Chapter 4, and the data quality objectives (DQOs) applicable to the particular investigation in question.

Details of the laboratory analyses must also be documented and closely followed. These should be compiled and presented in a standard operating procedure (SOP). Detailed analysis plans should be constructed for each instrument and analyte.

To produce high-quality data, a laboratory should have effective procedures in place for each of the following elements:

- bottle cleaning
- sample processing
- chemical analysis

Each of these elements is described in the sections that follow. In addition, the laboratory should follow effective procedures for documentation of method implementation and method changes. This chapter provides information on all major aspects of successfully operating a low-nutrient, low-ionic-strength water analysis laboratory.

## 3.2 LABORATORY PREPARATION PRIOR TO SAMPLE ANALYSIS

### 3.2.1 Bottle Cleaning

All bottles used for sample collection and for partitioning the sample into aliquots for transport to the laboratory must be clean and free of any contamination. Generally, it is the responsibility of the laboratory to provide clean bottles to the field crew in advance of initiating field sampling. Low-nutrient, low-ionic-strength samples can be more easily contaminated by improperly washed bottles than most other types of water samples because their low concentrations can be measurably altered by trace amounts of contaminants. Therefore, rigorous cleaning procedures must be used, which are often specific to the intended use of each size and type of bottle. Laboratories experienced with low-nutrient, low-ionic-strength surface water samples have adopted various methods for cleaning laboratory plasticware and glassware. Dilute acid (usually 1% or 2%) washing is often preferred for some analytes to solubilize potential contaminants that can then be removed with multiple deionized water (DIW) rinses. If acid washing is done, it should be carried out in a separate room with a negative-pressure ventilation system.

Rinse water should meet the specifications of the American Society for Testing and Materials for type III water (http://www.astm.org/Standard/standards-and-publications.html). A typical water deionization system that produces type III water includes, in sequence, a carbon removal tank, 1-$\mu$m filtration, and two mixed-bed cation-anion removal tanks. To ensure consistent water quality, specific conductance (or resistance) of the water should be monitored between the two primary treatment modules, in this case the mixed-bed tanks. If the specific conductance exceeds the preset limit of 1.0 $\mu$S/cm, an indicator light (which is normally on) is deactivated. The system is checked daily, and if the indicator light is off, the first tank is removed and replaced by the second tank. A new tank is then placed in the second position. By monitoring between the two tanks, the initial tank serves as a first-level treatment and the second tank serves as a polisher. By switching the polisher to the first position when the first-level tank no longer meets the standard, you extend the life span of the tanks and, most importantly, prevent substandard water from being used. These tanks degrade gradually, so the two-stage approach is needed to maintain a consistent level of water quality. The filter is replaced with every tank change, and the carbon tank is replaced depending on the volume of water treated and the organic carbon concentration of the influent. Once every 6 months is a typical replacement frequency. Note that appropriate methods for tank replacement are to some degree equipment specific. The procedure described here is one possible approach to producing thoroughly rinsed bottles. Other procedures may also be acceptable.

To ensure that the wash acid does not itself become a contaminant, repeated rinsing is followed by leaching the bottle with DIW. This is done by filling the rinsed bottle with DIW and storing it for 24 h or longer (Table 3.1). This step is necessary because acid and other contaminants (such as those from a previous sample) can migrate into the plastic matrix of the bottle wall. Over time, the contaminants can slowly leach out and change the sample concentration. The level of contamination caused by this process is usually low but can be sufficient to cause measurable increases in low-nutrient, low-ionic-strength samples. Simple rinsing does not necessarily eliminate this type of contamination.

Finally, each bottle is filled with DIW before shipping the bottles to the project location. Measurement of specific conductance of DIW stored in sample bottles, with an acceptance criterion of 1.2 $\mu$S/cm or less, provides QA for the bottle-cleaning procedure. We recommend using an acceptance criterion of 1.2 $\mu$S/cm or less, but we recognize that some laboratories use a somewhat higher criterion, as high as about 2.0 $\mu$S/cm. Specific conductance testing can be done on selected sample bottles that are treated as sample blanks.

Aliquot bottles that have had acid added for sample preservation are more likely to have contamination from bottle leaching than those that are

**TABLE 3.1    SUMMARY OF A TYPICAL BOTTLE-RINSING PROTOCOL FOR LOW-NUTRIENT, LOW-IONIC STRENGTH SAMPLES**

| Bottle | Acid Wash? | Rinses | DIW Soak Period |
|---|---|---|---|
| 60-ml plastic (cations, acidified) | Yes | 4 | 1 wk, 2X rinse, 24-h soak |
| 30-ml plastic (anions) | No | 6 | 24 h |
| 250-ml plastic (pH, ANC) | Yes | 4 | 24 h |
| 30-ml plastic (total Al, acidified) | Yes | 4 | 1 wk, 2X rinse, 24-h soak |
| 30-ml plastic (NH$_4$ or DON) | No | 6 | 24-hr |
| 40-ml glass (DOC) | Yes | 4 | 24-hr |
| DOC caps and liners | Yes | 6 | 24-hr |
| 1-L plastic (field, manual collection) | Yes | 4 | 24-hr |
| 500-ml plastic (field, autosample) | Yes | 4 | 24-hr |
| 1-L plastic (field, autosample) | Yes | 4 | 24-hr |

in contact with wash acid for 30 min or less. Therefore, preserved sample aliquot bottles require longer periods of DIW leaching. Plastic bottles should not be used if the caps have liners because the liners can become a source of contamination. Glass bottles, such as those needed for dissolved organic carbon (DOC) aliquots, usually have caps with removable plastic liners. The plastic liners must be removed from the caps for washing and be soaked in DIW for the same length of time that the bottle is soaked before being replaced in the cap.

Because the required cleaning procedure depends on the specific use of the bottle, a set of cleaning procedures needs to be developed and documented by the laboratory. Appropriate procedures to clean the various types of collection and aliquot bottles are listed here and summarized in Table 3.1. Although different laboratories may use different or modified procedures, it is critical to document the specific procedures used and to provide assurance that sample bottle contamination has been avoided.

Additional precautions may be needed when washing with acid solutions, such as hydrochloric acid (HCl). Local regulations may require separate disposal of acid (or basic) waste products (depending on the concentration), rather than pouring them down the sink. The possible need for special waste disposal procedures may affect the laboratory budget and may influence decisions regarding bottle-washing procedures.

### 3.2.1.1  Cleaning Bottles Used for Sample Collection in the Field

1. Remove sediment and other particles from the bottle with tap water, using a soft plastic brush if necessary, then rinse once with DIW. Note that use of a stiff bottle brush can scour the inside of the bottle, allowing contaminants to more easily adsorb to the bottle wall.
2. Fill with 2% HCl and let stand for 15 to 30 min in a separate bottle-washing room, away from analytical instrumentation. The need to use HCl in a separate bottle-washing room, away from analytical instrumentation, is because HCl can become volatilized and thereby contaminate nearby samples and, eventually, damage equipment.
3. Pour out the HCl and rinse thoroughly four times with DIW.
4. Fill with DIW and store for at least 24 h.
5. To prepare for transport to the field, empty the bottle and rinse once with DIW, then fill with DIW for transport to the field.
6. Before filling the bottle, the bottle should subsequently be rinsed three times in the field with the sample that is being collected.

### 3.2.1.2  Cleaning Bottles Used for General Laboratory Use, Unacidified Aliquots

1. Empty any remaining sample.
2. Fill with 2% HCl for 15 to 30 min.
3. Pour out the HCl and rinse thoroughly four times with DIW. (HCl can be reused for multiple washings but should be replaced and properly disposed of when it becomes discolored.)
4. Fill with DIW and store for at least 1 week.
5. To prepare for use, empty and then rinse twice with DIW.

### 3.2.1.3  Cleaning Bottles Used for General Laboratory Use, Acidified Aliquots

1. Pour out any remaining sample.
2. Rinse four times with DIW.
3. Fill with 2% HCl for 15 to 30 min.
4. Pour out the HCl and rinse thoroughly four times with DIW.
5. Fill with DIW and store for at least 24 h.
6. To prepare for use, empty and rinse once with DIW.

Some labs avoid the acid wash step for aliquots to be analyzed for $Cl^-$, $NH_4^+$, or dissolved organic nitrogen (DON), and instead rely on DIW leaching to remove any contaminants from the bottles. Depending on the intended use of sample aliquot bottles, the recommended number of rinses can vary. These differences are summarized in Table 3.1. As indicated, specific conductance should be measured for DIW stored in sample bottles as a QA measure. Such analyses should be conducted, at a minimum, on a subset of bottles prior to field use and on sample blanks during laboratory analysis.

## 3.2.2  Sample Processing, Preservation, and Storage

When samples arrive at the laboratory, they need to be accompanied by proper documentation using a chain-of-custody form. An example is provided in Appendix D. This form provides field information that includes project identification; when, where, how, and by whom the sample was collected; and information on the chain of custody that was followed. This form will need to have been checked by the field sampler against the information written on the sample bottle label to ensure that the information matches. We suggest a format for recording sample information in the field and sample transfer to the laboratory. The transfer of custody from field personnel to lab personnel must be documented by dated signatures on a chain-of-custody form. A copy of the signed form should be kept by both project and laboratory personnel. This procedure is needed to ensure that samples were collected and transferred to the laboratory as intended. Samples can be misplaced before arriving at the laboratory or within the laboratory before processing, particularly if there was a sample labeling error.

Prior to the start of sample processing in the laboratory, a unique code or sample serial number (SSN) is typically assigned by the laboratory and added to the laboratory data sheet. A single person (plus a trained backup) is generally assigned this responsibility to ensure that an SSN is not accidently used for more than one sample. If the SSN is assigned in the field, it is important that it is unique. The SSN will be used by the laboratory to track the sample through the steps of sample processing, chemical analysis, and data management. The information on the chain-of-custody form and laboratory data sheet will be entered into an electronic database. A variety of database software is available commercially, and some laboratories develop their own database system.

Each sample will typically be analyzed for a variety of constituents with specific processing needs that are chosen to meet stated project goals. Different results may be obtained from the same analysis if samples are prepared for analysis using different procedures. For example, results may differ if samples are filtered with filters of different pore sizes or of different materials. Sample processing can involve both preparation for analysis and preservation of the sample; therefore, it varies among analyses. The details of processing and analyses must be established to ensure that both sample processing and analyses done by the laboratory meet project needs.

Processing is generally accomplished by dividing the sample into several aliquots, each with its own filtration/no filtration, preservation, storage, and process and handling time requirements. For example, analysis for concentrations of base cations (BCs) (Ca, Mg, Na, and K) may require filtration through a 0.45-μm polycarbonate filter, whereas analysis of DOC may require filtration through a glass fiber filter to avoid possible organic contamination from the polycarbonate filter. Preservation of samples for analysis of BC and other metals that could form precipitates at nonacidic pH values usually involves the addition of nitric acid. However, addition of chemicals to samples should be avoided unless necessary to reduce the potential for sample contamination or alteration.

All samples that require filtering to remove particulate matter are normally filtered as soon as possible after arrival at the laboratory. Prompt filtration after sample collection is normally done for the purpose of removing bacteria, which can alter sample chemistry through their metabolic processes. Samples should be chilled as soon as possible after collection, up to the time of processing and preservation, because refrigeration retards microbial activity. Some aliquots will continue to be refrigerated until analysis (Table 3.2). Freezing is not necessary for most analytes but is recommended for DON and $NH_4^+$. The containers used to store aliquots prior to analysis also vary by analyte.

**TABLE 3.2    EXAMPLE LABORATORY ALIQUOT SCHEDULE FOR A PARTICULAR PROJECT**

| Aliquot[a] | Container | Filter | Treatment | Storage |
|---|---|---|---|---|
| A | 250-ml polyethylene | None | None | Refrigerator (4°C) |
| B | 30-ml polyethylene | 0.45-μm polycarbonate | None | Refrigerator (4°C) |
| C | 40-ml glass[b] | Glass fiber filter (GFF) | None | Refrigerator (4°C) |
| D | 60-ml polyethylene | 0.45-μm polycarbonate | 0.3 μl $HNO_3$ | Room temperature |
| E | 30-ml polyethylene (taped) | GFF: fill 2/3 full | None | Freezer (label DON) |
| F | 30-ml polyethylene (taped) | None: fill 2/3 full | None | Freezer (label $NH_4$)[c] |

[a] A—pH, ANC, specific conductance, total monomeric Al, organic monomeric Al; B—sulfate, nitrate, and chloride; C—dissolved organic carbon; D—calcium, magnesium, sodium, potassium, silicon; E—dissolved organic nitrogen; F—ammonium.

[b] Use of glass for storing samples in the laboratory prior to DOC analysis is preferred, but not essential.

[c] Ammonium concentrations (although typically low in natural waters) are unstable. The sample should be analyzed immediately on arrival at the laboratory or frozen until time of analysis.

Aliquots for BC analysis are generally stored in polyethylene or polypropylene bottles, which are economical and considered sufficiently inert with respect to BC, whereas aliquots for DOC analysis are usually stored in glass bottles to avoid organic contamination from plastic. An example of a typical sample-processing schedule for low-nutrient, low-ionic-strength samples is shown in Table 3.2. It is important to set up this type of schedule with laboratory person-nel to ensure that all samples from a particular project will receive the correct processing. In this example, reminders are included to tape bottle caps and not fill bottles completely for aliquots that are preserved by freezing.

Recommended laboratory holding times are given in Table 3.3. Those labora-tory holding times should be considered as guidelines or targets. Measurement of a sample analyte past the holding time is no justification for excluding that

**TABLE 3.3    RECOMMENDED LABORATORY HOLDING TIMES**

| Constituent | Holding Time |
|---|---|
| pH | 2 weeks |
| Conductivity | 2 weeks |
| ANC | 2 weeks |
| $NH_4$ | 3 months[a] |
| Dissolved N | 3 months[a] |
| $Al_m$ | 2 weeks |
| $Al_o$ | 2 weeks |
| Ca | 6 months |
| Mg | 6 months |
| Si | 6 months |
| Na | 6 months |
| K | 6 months |
| Cl | 1 month |
| $NO_3$ | 1 month |
| $SO_4$ | 1 month |
| DOC | 2 weeks |
| Turbidity | 2 days |

Source: Greg Lawrence, USGS Troy Laboratory, personal written communication, 2012.

[a] Samples for $NH_4^+$ and dissolved N are preserved by freezing and analyzed in batches.

concentration value from the database. Nevertheless, sample measurements taken beyond the specified holding time should be flagged.

Each laboratory should have detailed documentation of the steps used in sample processing. An example list of sample-processing steps follows:

1. Obtain chain-of-custody forms that have assigned SSNs and aliquot labels that correspond to the processing selected for the project. Initiate a laboratory data sheet.
2. Retrieve clean aliquot containers and place appropriate numbered dots or label tape on them. Put on gloves, empty the containers, and rinse them with DIW, if applicable.
3. Retrieve the field samples to be filtered from the refrigerator. If there is insufficient sample volume to prepare all of the required aliquots, be sure to follow the procedure for low-volume samples.
4. Shake the field sample bottle. Rinse the aliquot bottles with a small volume of the sample. Fill aliquot containers with raw sample for the aliquots that do not require filtering. Fill in the letter code that corresponds to the aliquots on the laboratory data sheet to document the processing method.
5. Retrieve the filtering apparatus from the DIW soak and rinse it well with DIW. Set up the filtering apparatus on the vacuum manifold.
6. Place an appropriate filter (handling the edge of the filter only) on the apparatus with tweezers that have been rinsed with DIW. For 0.4- and 0.1-μm filters, place the shiny side up where appropriate (filters are sometimes packaged shiny side down). Filter 10 ml of DIW, then rinse the inside of the chamber with 10 ml of sample and filter into a waste container. Discard filtrate.
7. Place proper aliquot container under filtering apparatus. Filter 5–10 ml of sample into container, rinse, and then discard the filtrate.
8. Filter an appropriate amount of sample into container. If another aliquot of the same sample requires the same filter, repeat starting at step 7. Fill in the letter code that corresponds to the aliquots on the laboratory data sheet. Discard the used filter and rinse the filtering apparatus with DIW.
9. Repeat steps 6 through 8 for each sample aliquot. If the filter clogs, replace with a new filter following step 6, then go to step 8.
10. After samples have been processed, rinse the filtering chambers with DIW and place them in DIW soak buckets. Replace the DIW in the buckets weekly.
11. Store the remaining sample volume for possible reanalysis, at least until QA/QC analyses have been completed. Once it has been determined that the analysis meets DQOs, discard the remaining sample in field sample bottles and bring the bottles to the bottle-washing room.

**TABLE 3.4    EXAMPLE LOW-VOLUME SAMPLE SCHEDULE**

| Priority | Aliquot Type[a] |
|---|---|
| 1 | B, 15 ml |
| 2 | D, 15 ml |
| 3 | A, 50 ml |
| 4 | C, 15 ml |
| 5 | H, 15 ml |
| 6 | G, 20 ml |

[a] See Table 3.2 for the description of aliquot types. Fill aliquot bottles from the available sample volume to the appropriate aliquot volume in the listed order of priority.

12. Aliquots that require acidification should be acidified in the hood using the appropriate acid dispenser.
13. Aliquots should be stored in the appropriate places as described in the specific project sample-processing schedule.
14. Date and processor's initials must be recorded on the laboratory data sheet. Completed chain-of-custody forms and laboratory data sheets should be filed in a safe location. Information on these forms will need to be entered into the electronic database.

In some situations, it is possible that the sample bottle was not completely filled in the field. If there is insufficient sample volume for all aliquots, analysis of the sample volume available must be prioritized based on the objectives of the project. If there is an anticipation that some low-volume samples will be collected, a low-volume schedule needs to be prepared and made available to laboratory personnel in advance of sample arrival at the laboratory. An example of a low-volume schedule is provided in Table 3.4.

## 3.3   CHEMICAL ANALYSIS

The constituents that need to be measured in a water sample will be determined by the specific objectives of the project. Each method of chemical analysis will provide a constituent concentration with a certain level of accuracy and precision over a finite concentration range that is specific to that method. Low-nutrient, low-ionic-strength waters generally require methods that are effective at the lowest concentration ranges. A variety of methods are usually available to determine the concentration of a given constituent, even at low-concentration ranges. The method selected must (1) be appropriate for the expected concentration range; (2) provide the data with the accuracy and precision to successfully achieve the DQOs specified in the QA plan; and (3) not exceed logistical limitations with regard to sample collection, sample preparation, or laboratory capabilities. For example, a method for determining $NH_4^+$ concentrations might provide data over the necessary concentration range with a sufficiently high level of accuracy and precision but may require that the analysis be done within 12 h of collection. Such a short holding time might not be feasible for samples collected from remote sites or might be beyond the capabilities of the laboratory.

Selection of a method that is capable of meeting the required data accuracy and precision specified in the DQOs does not ensure that this level of data quality will be achieved. Rigorous QC and QA procedures must be followed as part of the method implementation. QC refers to procedures that identify results during chemical analysis that do not meet DQOs, thereby triggering immediate corrective action that usually involves reanalysis of that sample. DQOs are generally based on (1) the precision and accuracy levels required by the project and the laboratory and (2) the analytical limits of the methods used. A key component of QC is the introduction of artificial samples of known concentration, which are associated with a specific set of project samples. QC procedures are generally focused on instrument performance.

Additional procedures, referred to as QA, which are also evaluated by DQOs, are used to document laboratory performance through the introduction of artificial and natural samples that are not associated with a specific set of project samples, but reflect the accuracy and precision of sample preparation and analysis, including instrument performance. Recommended protocols for QA and QC are described in Chapter 4 of this book.

Each method used to determine a chemical concentration involves a complex set of procedures, reagents, and instrumentation. Any variation in these factors can potentially change the result, yielding a different concentration value. Therefore, each method requires an SOP that is strictly adhered to by the analyst each time that the method is implemented. All details of the method must be documented in the SOP, which must be available to any potential user of the data. Each SOP must be dated, authored, and approved. Any change to an SOP requires that the date of preparation also be changed on the SOP document.

We have not attempted here to recommend specific SOPs for implementation of analysis methods. SOPs are specific to individual laboratories and instrumentation and can be influenced by the expected concentrations in a particular study. Analytical methods, and thus SOPs, will inevitably change as instrumentation and technology improve. Moreover, as indicated, the specific analyses, methods, and details of SOPs must complement the project objectives and DQOs. Rather than recommending specific SOPs for general use, we recommend that each project, in consultation with the analytical laboratory, should adopt and adhere to SOPs that ensure and document attainment of appropriate DQOs through all phases of data acquisition, including sample collection, handling, analysis, and reporting. Selection of DQOs and development of QA plans are discussed in Chapter 4 of this book.

Detailed methods such as these should be prepared by the laboratory, in advance, for each parameter to be measured in the water quality study. Alternative SOPs may be used as long as QA/QC objectives are satisfied.

# Chapter 4

## Quality Assurance/Quality Control

### 4.1 INTRODUCTION

After water samples have been collected and analyzed in the laboratory, a project database can be compiled. But, that database is not ready for statistical and graphical analysis and interpretation until it has been quality assured. Mistakes and errors of a variety of kinds can crop up at many stages of project implementation. Such errors can compromise the quality of the data. You have devoted considerable effort and expense in generating chemical concentration values (CVs). You now need to know how robust those numbers are. If any of the data have been compromised in any way, you need to know that before you move forward with data analysis and interpretation. You risk faulty conclusions regarding patterns or trends in the data. Furthermore, these data may be needed as the basis for management or policy decisions. They may be used in litigation focused on emissions quantities, legal limits, and assignment of blame for environmental pollution. For a variety of reasons, you do not want and cannot afford bad data. You have too much at stake. You need to know yourself that the data are of high quality, and you also might need to be able to demonstrate to others (scientists, stakeholders, resource managers, lawyers, judges) that the data are of high quality. Often, you do not know in advance how the data will be used. You may conduct a study that seems at first glance to be mostly of interest only to you. But, years later, your study may shed needed light on a problem that is the subject of a lawsuit or that is needed as the basis for important management decisions.

So, how do you ascertain the quality of the data? You implement a quality assurance/quality control (QA/QC) program, which impinges on virtually all components of the project, including sample collection, laboratory analysis, data reporting, and data analysis. Elements of the QA/QC program are described here.

Selection of the water quality constituents that you measure in the laboratory will be determined by the specific objectives of the project. Therefore, one of the first steps in project planning is to determine which variables you want to measure. Often, your wish list will have to be pared down to fit your available budget. You will need to balance competing desires to sample at more sites and on more sample occasions and to measure more variables. If you relax your requirements in one area, you will be able to increase your level of effort in another area for the same available budget. Such trade-offs constitute a normal part of the project scoping effort. Also, in considering your available budget, remember that you will need to increase your total sample allocation by about 10–20% to accommodate laboratory analysis of QA samples.

Each method of chemical analysis will provide a constituent concentration with a certain level of accuracy and precision over a finite concentration range that is specific to that method. Low-nutrient, low-ionic-strength waters, such as those commonly included in acidic deposition or nutrient enrichment studies, generally require methods that are effective at the lowest concentration ranges. A variety of methods are usually available to determine the concentration of a given constituent. The method selected must (1) be appropriate for the expected concentration range in the water bodies of interest, (2) provide data with the accuracy and precision to successfully achieve project objectives, and (3) not exceed logistical limitations with regard to sample collection, sample preparation, or laboratory capabilities. For example, a method for determining $NH_4^+$ concentration might provide data over the necessary concentration range with a sufficiently high level of accuracy and precision but may require that the laboratory analysis be conducted within 12 h of sample collection. Such a short holding time might not be feasible for remote sites or might be beyond the capabilities of the laboratory.

A variety of laboratory methods and instruments might be capable of providing data of suitable quality for a particular study. Nevertheless, selection of a method that is capable of meeting the required data accuracy and precision does not ensure that this level of data quality will be achieved within a given project. You must demonstrate that your selected methods perform as expected. Rigorous QC and QA procedures must be followed as part of the method implementation. Both QA and QC procedures are important parts of any sampling program, whether field or laboratory. Application of QA/QC procedures will provide the basis for determining the quality of the resulting data. In the absence of appropriate QA/QC procedures, it is impossible to judge whether the data are adequate to meet the needs of the project.

Here, QC refers to procedures that identify measurements during the chemical analysis that do not meet data quality objectives (DQOs), thereby triggering immediate corrective action that usually involves reanalysis of that sample in the laboratory. DQOs are generally based on (1) the precision and accuracy levels required by the project and the laboratory and (2) the analytical limits of the methods used. QC procedures are primarily focused on instrument performance. A key component of QC is the analysis of synthetic samples of known concentration that are analyzed along with a set of project samples.

Additional procedures, referred to as QA, are used to document laboratory performance through the introduction of artificial and natural samples that are not associated with a specific set of project samples but reflect the accuracy and precision of sample preparation and analysis, including instrument performance.

There are three primary components to QA for the project laboratory:

1. Routine evaluation of laboratory analytical performance related to adopted DQOs
2. Strict adherence to project standard operating procedures (SOPs), including sample bottle preparation, sample collection, sample processing, and analysis methods
3. Submission of measurement data QA results along with reported analytical data

The various attributes of data quality and how they are evaluated are described next.

## 4.2  ATTRIBUTES OF DATA QUALITY

The goal of any field research project is to produce sound analyses and high-quality data. Establishment of DQOs and development of a QA plan are important to ensure that data meet the established objectives for precision, accuracy, representativeness, completeness, and comparability. Each type of QC/QA sample or process is generally associated with a DQO. The value of the DQO for each analyte will be set by project objectives and is usually method specific. When the range of acceptable values measured for a sample of known concentration and defined by a DQO is exceeded, the method is considered to be out of control limits, and remedial action must be taken in the laboratory.

### 4.2.1  Method Detection and Reporting Limits

#### 4.2.1.1  Method Detection Limit

For chemical measurements, requirements for the method detection limit (MDL) need to be established. The term *detection limit* has been used in various ways when referring to the lower limit of a method concentration range. This lower limit can be a function of instrument capability, chemical reactions that are part of the method, or both. The most basic definition of a detection limit is the threshold below which measured values are not considered statistically different from a blank value (Helsel 2005). Blank values are measurements of samples of deionized water (DIW; water with no other ions in it). Thus, measured concentrations below the detection limit are not statistically different from zero. The repeated measurement of a sample with a known concentration that is at or near the detection limit will exhibit considerable variability that is assumed to be normally distributed. Therefore, in this range of measurement, separating a true CV from a value resulting from analytical noise is problematic. This is of particular consequence for research or monitoring objectives that involve the detection of trace contaminants such as Hg or organic contaminants such as pesticides and low levels of nutrients or other analytes in dilute waters.

The MDL is defined as the lowest level of analyte that can be distinguished from zero. The first step in determining the detection limit is to make your best estimate of the value of the MDL. A set of standards is then made with sequentially decreasing concentrations that extend above and below the initial estimate of the MDL. Each concentration should be analyzed seven times to provide a mean and standard deviation for each CV (Helsel 2005). The true MDL will occur at the concentration that is not statistically different from the next-lowest concentration. This is determined by running $t$ tests between the paired concentrations, starting with the two highest concentrations. Then, you run a $t$ test between the second-highest concentration and the third-highest concentration, working downward until you reach a pair of concentrations that are not statistically different. For example, you might find that the fourth- and fifth-lowest concentrations were not statistically different ($p < .05$). In other words, the method could not detect the difference between these two concentrations. The third-lowest concentration would then be the true MDL. Determination of the MDL is demonstrated in the following example:

Step 1. Assume that your best estimate of the MDL for a particular method equals 0.01 µg/L.

Step 2. Make up a set of solutions with the following concentrations that bracket the estimated MDL, numbered from highest concentration to lowest concentration:

1. 0.06 µg/L
2. 0.04 µg/L
3. 0.02 µg/L
4. 0.01 µg/L
5. 0.005 µg/L
6. 0.003 µg/L

Step 3. Analyze each solution seven times, then calculate a mean and standard deviation from the seven measured values obtained for each solution. Your mean values might look like the following:
1. 0.055 μg/L
2. 0.039 μg/L
3. 0.022 μg/L
4. 0.011 μg/L
5. 0.014 μg/L
6. 0.013 μg/L

Step 4. Then, run $t$ tests sequentially between each pair of concentrations (1 versus 2, 2 versus 3, 3 versus 4, etc.).

Step 5. You find that results for solutions 1 versus 2, 2 versus 3, and 3 versus 4 are statistically different, but results for solutions 4 versus 5 are not statistically different. This means that the concentration of solution 3 is your MDL (0.02 μg/L).

A variety of less-rigorous methods of determining detection limits is also used. One of the most common methods determines the MDL by multiplying the standard deviation of repeated measurements of the estimated MDL by a factor of three. The accuracy of the detection limit determined in this manner will depend on the accuracy of your estimated MDL, which is unknown. Therefore, we recommend the stepwise determination as described.

The MDL can vary from run to run, and over time, in response to such issues as change in analyst, new instrumentation, or the aging of instrumentation. Therefore, the initial analysis to determine the MDL should be repeated three times over several weeks and thereafter at least annually for constituents with concentrations in water samples that commonly occur near or below the MDL. Laboratories that focus on the measurement of trace contaminants may determine MDLs more frequently, but otherwise, some unmeasured variation in the MDL will not negatively affect data quality. If a new analyst is appointed or equipment is replaced, a new MDL value should be determined, regardless of the length of time since the last MDL was determined.

The MDL is not to be confused with the upper limit of the concentration range of a particular method. At concentrations above the method range, the relationship between measurements of standards and the known concentrations of these standards can change, thereby requiring a different standard curve.

Because measurements that fall below the MDL are statistically indistinguishable from zero, they are often set to zero in the database. For example, this is the recommendation of Helsel (2005). In any case, the MDL values associated with analysis results should be made available to data users. The topic of censoring data based on MDL values is discussed further in Section 5.3.1.

### 4.2.1.2   Reporting Limit

The reliability of measurements that are above, but near, the detection limit is less than the reliability of measurements at higher concentrations. The concentration above which measurement variability becomes acceptably low defines the threshold referred to as the reporting limit. Like MDLs, reporting limits are low concentrations, but they are always higher than the MDL, at least by a small amount. At measured concentrations above the reporting limit, the method is considered reliable and therefore subject to DQOs established for precision and accuracy.

Measurements that fall in the narrow range below the reporting limit, but above the MDL, may not consistently meet the DQOs for reproducibility or accuracy and should be flagged in the database. We recommend retaining these values in the database because they indicate low, nonzero concentrations and therefore provide information that could be useful. The flag should caution the user that the precision and accuracy of these low measured values are uncertain and likely to be higher than measurements that fall above the reporting limit.

Reporting limits are determined for each chemical analysis by establishing the precision and accuracy of measurements in the lower portion of the method concentration range. To determine the reporting limit for a particular analysis, the steps outlined next should be followed:

Step 1. Select a relative DQO (±%) for both precision and accuracy. A value of 10% is commonly used for most analytes.

Step 2. Make an estimate of the concentration that defines the reporting limit. The values for reporting limits listed in Table 4.1 can be used to provide these estimates. From Table 4.1, for example, the estimated reporting limit for $NO_3^-$ analysis is 0.1 mg/L.

Step 3. Make up three solutions of known concentrations that are higher than the estimated reporting limit, one solution with a concentration equal to the estimated reporting limit, and three solutions of known concentrations that are lower than the estimated reporting limit but higher than the MDL (listed as 0.03 mg/L for $NO_3^-$ in Table 4.1). For example, the $NO_3^-$ concentrations in the test solutions might be as follows:
   1.   0.6 mg/L
   2.   0.4 mg/L
   3.   0.2 mg/L
   4.   0.1 mg/L
   5.   0.08 mg/L
   6.   0.06 mg/L
   7.   0.04 mg/L

Step 4. Analyze each solution five times and calculate the mean and standard deviation of the five values at each concentration level.

**TABLE 4.1 RECOMMENDED DQOs FOR DETECTION LIMITS, PRECISION, ACCURACY, REPORTING LIMIT, AND COMPLETENESS[a]**

| Variable or Measurement | Method Detection Limit | DQOs for Precision and Accuracy | | Reporting Limit | Completeness |
|---|---|---|---|---|---|
| | | Relative DQOs (± %) | Absolute DQOs (± Concentration Value) | | |
| Oxygen, dissolved | NA | NA | 0.5 mg/L | NA | 95% |
| Temperature | NA | NA | 1°C | NA | 95% |
| pH, closed system and equilibrated | NA | NA | 0.15 pH units | None | 95% |
| Acid-neutralizing capacity | NA | 15% | 6 µeq/L | None | 95% |
| Carbon, dissolved inorganic, closed system | 0.10 mg/L | 10% | 0.1 mg/L | 0.5 mg/L | 95% |
| Carbon, dissolved organic | 0.1 mg/L | 10% | 0.1 mg/L | 0.5 mg/L | 95% |
| Conductance | NA | 10% | 1 µS/cm | None | 95% |
| Aluminum, total dissolved, total monomeric, and organic monomeric | 10 µg/L | 10% | 0.02 mg/L | 27 µg/L | 95% |
| *Major cations:* | | | | | |
| Calcium | 0.02 mg/L | 10% | 0.02 mg/L | 0.08 mg/L | |
| Magnesium | 0.01 mg/L | 10% | 0.02 mg/L | 0.02 mg/L | |
| Sodium | 0.02 mg/L | 10% | 0.02 mg/L | 0.03 mg/L | |
| Potassium | 0.04 mg/L | 10% | 0.04 mg/L | 0.05 mg/L | |

*Continued*

**TABLE 4.1 (*Continued*)   RECOMMENDED DQOs FOR DETECTION LIMITS, PRECISION, ACCURACY, REPORTING LIMIT, AND COMPLETENESS[a]**

| Variable or Measurement | Method Detection Limit | DQOs for Precision and Accuracy | | Reporting Limit | Completeness |
|---|---|---|---|---|---|
| | | Relative DQOs (±%) | Absolute DQOs (±Concentration Value) | | |
| Ammonium | 0.02 mg/L | 10% | 0.02 mg/L | 0.04 mg/L | 95% |
| *Major anions:* | | | | | 95% |
| Chloride | 0.03 mg/L | 10% | 0.03 mg/L | 0.1 mg/L | |
| Nitrate | 0.03 mg/L | 10% | 0.03 mg/L | 0.1 mg/L | |
| Sulfate | 0.05 mg/L | 10% | 0.05 mg/L | 0.2 mg/L | |
| Silica | 0.05 mg/L | 10% | 0.05 mg/L | 0.4 mg/L | 95% |
| Phosphorus, total | 1 µg/L | 10% | 0.002 mg/L | 4 µg/L | 95% |
| Nitrogen, total | 0.07 mg/L | 10% | 0.03 mg/L | 0.15 mg/L | 95% |
| True color | NA | 10% | 5 PCU[b] | None | 95% |
| Turbidity | NA | 10% | 2 NTU[c] | None | 95% |
| Total suspended solids | 0.1 mg | 10% | 1 mg/L | 0.4 mg/L | 95% |

*Note:* NA, not applicable.

[a] DQOs for precision and accuracy are expressed two ways: in relative terms (±percentage measured value) and in absolute terms (± actual measured concentration). The DQO is considered to be met if *either* of these criteria is satisfied.

[b] platinum cobalt units

[c] nephelometric turbidity units

Step 5. Using the mean values, calculate the accuracy (expressed as percentage error) and precision (expressed as the coefficient of variation) following the procedures given in the next section. Your data may look something like this:

1.  0.6 mg/L;    % error = 5.4    Precision = 6.8
2.  0.4 mg/L;    % error = 6.2    Precision = 4.9
3.  0.2 mg/L;    % error = 7.5    Precision = 6.3
4.  0.1 mg/L;    % error = 7.2    Precision = 7.5
5.  0.08 mg/L;   % error = 8.1    Precision = 9.5
6.  0.06 mg/L;   % error = 21.8   Precision = 25.0
7.  0.04 mg/L;   % error = 45.6   Precision = 39.3

Step 6. Determine the test concentration above which your measurements of error and precision are both less than or equal to 10%. For this example, that is a concentration of 0.06 mg/L. The next-highest test concentration in the series is then designated as the reporting limit. Based on these data and a DQO of 10% for both accuracy and precision, your reporting limit would be 0.08 mg/L.

## 4.2.2  Precision and Accuracy

Precision and accuracy are estimates of random and systematic error in a measurement process. Together, they provide an estimate of the total error or uncertainty associated with an individual measurement. Precision is measured by repeated analysis of a single sample. The variation of these measurements indicates the level of method precision. Accuracy is an indication of how closely the measurements match the true concentration of the sample. The distinction between precision and accuracy is shown in Figure 4.1.

Accuracy can be determined from measurements of solutions of known composition or from the analysis of samples that have been fortified by the

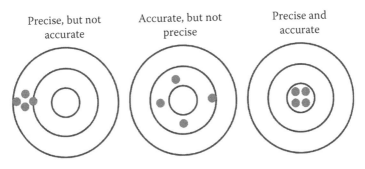

Precise, but not          Accurate, but not          Precise and
accurate                  precise                    accurate

**Figure 4.1**   Schematic illustration of precision and accuracy.

addition of a known quantity of analyte. Accuracy is quantified by relating the measured value of a QC sample to the known value of that QC sample. It is usually expressed as percentage error. For QC samples, the DQO objective is defined as the value of the percentage error. If the measured concentration is greater than the known value plus the DQO or less than the known value minus the DQO, the method is considered to be out of the control limits. The percentage error is calculated as follows:

$$\% \, error = \frac{known \, concentration - measured \, concentration}{known \, concentration} \times 100 \qquad (4.1)$$

Accuracy can also be quantified through analysis of interlaboratory reference samples. For example, the US Geological Survey (USGS) Standard Reference Program provides a most-probable value that can be used to calculate percentage error in the same manner that known values for QC samples are used. The same approach can be used with the Canadian National Water Research Institute (NWRI) Program, which also provides a most-probable value $D$ as follows:

$$D = [(AV - MCN)/MCV] \times 100 \qquad (4.2)$$

where $AV$ is the analyzed value, and $MCV$ is the mean CV (most-probable value for source material).

Method precision is evaluated by analyzing multiple, often duplicate or triplicate, project samples. Ideally, each time a sample is reanalyzed, the same CV should be reproduced. Precision is typically quantified by the CV. The DQO is defined as the CV above which the method is out of control. The CV is calculated as

$$CV = \frac{S}{\bar{X}}(100) \qquad (4.3)$$

where $S$ is the standard deviation, and $\bar{X}$ is the arithmetic mean of replicate samples. It should be noted that relative precision (e.g., CV) is not independent of concentration. For low concentrations, criteria for both bias and precision are typically expressed in terms of absolute rather than relative error.

## 4.2.3  Completeness

Completeness requirements are established and evaluated as the percentage of valid data obtained versus the amount of data expected. Thus, completeness quantifies the extent to which data are missing. Completeness objectives are usually designated as over 95% for each variable. If there are logistical problems associated with collection of some of the desired data, the completeness criterion might need to be relaxed.

## 4.2.4  Comparability

Comparability is defined as the confidence with which one data set can be compared to another. Comparability is enhanced by the use of standardized sampling procedures in the field and laboratory. Comparability of data is also facilitated by implementation of standardized QA and QC techniques. For all measurements, reporting units and formats are specified in advance and recorded on field forms and laboratory databases in these units and formats. Comparability is also addressed by providing QA data on detection, precision, and accuracy and by conducting methods comparison studies when necessary and participating in interlaboratory performance evaluation (PE) studies, such as those conducted by the USGS and NWRI of Canada. To provide estimates of trends in any analyte or indicator, data collected each year must be comparable to data collected in all prior and subsequent years. Comparability can be quantified through comparison of precision and accuracy estimates obtained from QA samples.

## 4.2.5  Representativeness

Representativeness is the degree to which the data accurately and precisely represent the environmental attribute of interest. Although representativeness is not a laboratory QA/QC issue, it is affected by problems in all other attributes of QA. A representative sample requires that the sample site be reflective of the study population of interest and that the sample itself is representative of the system of interest (e.g., that the water sample collected in the field reflects the condition in the subject lake or stream). Representativeness is ensured by following all field- and laboratory-sampling procedures and holding time requirements to ensure that analytical results are representative of the conditions at the time of sampling. Use of QA and QC samples similar to the type of environmental samples analyzed provides estimates of precision and bias that are applicable to the collected data.

Representativeness can also pertain to the extent to which the selected sampling sites capture the distribution of anticipated results across the population of interest. For example, a synoptic survey of streams within a given region may be designed to represent the population of streams of interest. These might be, for example, those streams in the study region above or below a certain stream order, within a specified range of watershed areas, on a specific bedrock lithology, or those represented as blue lines on a certain resolution topographic map. Typically, in the absence of a census of all streams in the population of interest, site representativeness will be achieved by means of a statistical (random) site selection. This can entail application of simple random sampling or stratified random sampling. The latter approach allows the project design team to purposely oversample or undersample certain types of water bodies and still maintain the ability to calculate population statistics, such as the number, percentage, or length of stream reaches that satisfy certain conditions. This might be, for example, streams having acid-neutralizing capacity (ANC) below 50 µeq/L, $NO_3^-$ above 20 µeq/L, or total P above 0.1 mg/L. A stratified random sampling may be useful when, for example, you do not want to spend a large percentage of your project resources on data collected at sites that are not highly sensitive to, or impacted by, the stressors under investigation. Rather, the majority of your sampling sites will be from the subpopulation of greatest interest, yet you retain the ability to make quantitative estimates for the entire population.

## 4.2.6  Recommended Laboratory Data Quality Objectives

Each laboratory must have its own set of DQOs that pertain to the quality of the analytical data that are routinely produced by the laboratory. Projects also have data quality requirements that are based on the objectives and resources of the specific project in question. Laboratory DQOs must be evaluated to ensure that the laboratory is capable of delivering the accuracy and precision that the project requires. In general, we recommend DQOs for detection, accuracy, and precision as specified in Table 4.1. These DQO values are used by the USGS, New York Water Science Center Water and Soil Analysis Laboratory, Troy, New York, which specializes in the analysis of low-ionic-strength waters for air pollution effects research projects (Greg Lawrence, USGS, personal written communication, 2010).

These recommended guidelines for precision and accuracy DQOs given in Table 4.1 may not be appropriate for all projects. You might determine, for a specific analyte and project, that one or more recommended guidelines can be relaxed, especially if the laboratory is unable to achieve the recommended level of data quality and if the project does not require such high levels of precision and accuracy. Conversely, you might determine that a particular project would require higher standards of precision and accuracy. In general, we believe that the values presented in Table 4.1 will satisfy the needs of most anticipated routine atmospheric deposition effects sampling projects.

As represented in Table 4.1, we recommend application of DQOs for precision and accuracy that are calculated two ways: based on relative percentage variation and based on absolute variation. A given DQO can be considered to be met if *either* of these two conditions is satisfied. In general, conformance with the DQO for accuracy and precision will be determined by evaluation of relative variation. However, at low CVs, the relative DQOs can be difficult or impossible to achieve. For example, if the ANC of a particular stream is 10 µeq/L, the relative DQO for precision and accuracy of the ANC measurement is 15% (Table 4.1), or 1.5 µeq/L. There is no laboratory that can achieve that level of accuracy and precision in measuring ANC. For a sample having such low ANC, however, the absolute DQO (6 µeq/L; Table 4.1) is considered to be achievable. As long as the absolute DQO criterion is satisfied, the DQO for precision and accuracy is considered to be met.

For most analytes, our recommended relative DQO for precision and accuracy is 10%. Nevertheless, we believe that most laboratories should be able to do better than that. A good target DQO in most cases is ±5%; this is the level of precision and accuracy that the laboratories and projects should strive to reach.

## 4.3   QA/QC SAMPLE TYPES

The following sections describe the various types of samples and DQOs that are typically used for QC and QA in laboratories that specialize in analysis of low-nutrient, low-ionic-strength waters. There is no definitive rule regarding how many QA/QC samples should be included in a given project. This will be determined, in part, by the intended use of the data and the available budget. In general, we recommend that at least 30% of the samples analyzed in the laboratory for a given project be QA or QC samples, distributed among the types of samples discussed in the sections that follow.

The QC samples are used to measure the accuracy of an instrument's calibration and to detect variations in instrument response within an analytical run. Types of laboratory QC samples are summarized in Table 4.2. These samples are made up in the laboratory using type I DIW and purchased chemicals. Source material for all QC samples is from either a manufacturer other than the producer of the source material used to make calibration standards or a lot other than the source material used to make calibration standards. QC-high and QC-low samples are analyzed within a given laboratory run for most constituents. Exceptions are ANC, pH, and specific conductance. Either the QC-high sample or the QC-low sample is analyzed within an ANC, pH, and specific conductance run, depending on the expected concentration range of the environmental samples. This reduces the chance of carryover from a low-pH (or low-ANC or low-specific-conductance) QC sample to a high-pH project sample through the transfer of the electrode between samples.

We recommend that QC samples are analyzed immediately after instrument calibration, once after every 10 project samples, and at the end of each run. QC samples that do not meet DQOs for accuracy are rerun. If the value is then

**TABLE 4.2    RECOMMENDED LABORATORY QUALITY CONTROL SAMPLES**

| QC Sample Type (Analyte and Description) | Frequency | Acceptance Criteria | Corrective Action |
|---|---|---|---|
| Laboratory blank: (all analyses except pH and TSS) Reagent blank: (DOC, Al [total, monomeric, and organic monomeric], ANC, $NH_4+$, $SiO_2$) | Once per batch prior to sample analysis | Control limits $< \pm$MDL or $< 1 \mu$M, whichever is least restrictive | Prepare and analyze new blank. Determine and correct problem (e.g., reagent contamination, instrument calibration, or contamination introduced during filtration) before proceeding with any sample analyses. Reestablish statistical control by analyzing three blank samples. |
| Filtration blank (all dissolved analytes, excluding syringe samples) ASTM type II reagent water processed through filtration unit | Prepare once per week and archive | Measured concentrations < MDL | Measure archived samples if review of other laboratory blank information suggests source of contamination is sample processing. |
| Detection limit quality control check sample (QCCS) (all analyses except true color, turbidity, and TSS): prepared so concentration is approximately four to six times the required MDL | Once per batch | Control limits $< \pm$MDL | Confirm achieved MDL by repeated analysis of appropriate standard solution. Evaluate affected samples for possible reanalysis. |
| Calibration quality control check sample (CQCCS)[a] | Before and after sample analyses | Control limits < precision objective: Mean value < bias objective | Repeat CQCCS analysis. Recalibrate and analyze CQCCS. Reanalyze all routine samples (including performance evaluation and field replicate samples) analyzed since the last acceptable CQCCS measurement. |

| | | | |
|---|---|---|---|
| Internal reference sample (suggested when available for a particular analyte) | One analysis in a minimum of five separate batches | Control limits < precision objective Mean value < bias objective | Analyze standard in next batch to confirm suspected imprecision or bias. Evaluate calibration and CQCCS solutions and standards for contamination and preparation error. Correct before any further analyses of routine samples are conducted. Reestablish control by three successive reference standard measurements that are acceptable. Qualify all sample batches analyzed since the last acceptable reference standard measurement for possible reanalysis. |
| Laboratory replicate sample (all analyses) For closed-headspace syringe samples, replicate sample represents second injection of sample from the sealed syringe | One per batch | Control limits < precision objective | If results are below MDL, Prepare and analyze split from different sample (volume permitting). Review precision of CQCCS measurements for batch. Check preparation of split sample. Qualify all samples in batch for possible reanalysis. |

*Continued*

**TABLE 4.2 (*Continued*)  RECOMMENDED LABORATORY QUALITY CONTROL SAMPLES**

| QC Sample Type (Analyte) and Description | Frequency | Acceptance Criteria | Corrective Action |
|---|---|---|---|
| Matrix spike samples (only prepared when samples with potential for matrix interferences are encountered) | One per batch | Control limits for recovery cannot exceed 100% ± 20% | Select two additional samples and prepare fortified subsamples. Reanalyze all suspected samples in batch by the method of standard additions. Prepare three subsamples (unfortified, fortified with solution approximately equal to the endogenous concentration, and fortified with solution approximately twice the endogenous concentration). |

*Note:*  ASTM, American Society for Testing and Materials; TSS, total suspended solids.

[a] For turbidity, a CQCCS is prepared at one level for routine analyses (US EPA 1987). Additional CQCCSs are prepared as needed for samples having estimated turbidities greater than 20 NTU. For TSS determinations, CQCCS is a standard weight having mass representative of samples.

acceptable, the run is continued. If the rerun QC sample value is unacceptable, the project sample data preceding it are considered out of control: The data are rejected, and the instrument is recalibrated. Only accepted QC sample and project sample data are entered into the database. The analytical results of QC samples should be recorded to indicate (1) the frequency of out-of-control data that are not rerun and (2) biases and trends of control data.

## 4.3.1  Filter Blanks, Analytical Blanks, and Field Blanks

Blanks are aliquots of DIW that are processed and analyzed in the same manner as project samples. Filter blanks are analyzed only for constituents that require filtration. Filter-blank analysis indicates whether detectable contamination has occurred during any step in sample handling that occurs in the laboratory, including bottle-washing procedures, filtration, sample preservation, and chemical analysis. Analytical blanks are aliquots of type I DIW that are processed and analyzed as project samples, except that the filtration step is omitted. Contamination of analytical blanks may be attributed to any step in sample handling other than filtration, including the quality of DIW. The use of an analytical blank together with a filter blank therefore enables contamination from filtration to be isolated from contamination during DIW preparation or other phases of sample preparation and analysis. The use of both a filter blank and an analytical blank is recommended because the filtration process poses the greatest single source of potential sample contamination. A filter blank and an analytical blank should be included as a QC pair in the sample stream at a frequency of at least 1 per 50 project samples.

Some programs require a QA sample referred to as a field blank. The field blank is prepared by bringing DIW into the field, then transferring it to a sample bottle or transporting it back to the analytical laboratory. From that point forward, the DIW in the sample bottle is treated as any other sample collected in the field. It is not clear what information this procedure provides because the action involved does not replicate any aspect of the actual field sampling. We therefore do not recommend the collection of field blanks unless some specific project objective requires field filtration of water samples, which could introduce the potential for sample contamination and which might be revealed in the process of transferring and filtering the blank water in the field.

## 4.3.2  Replicate Environmental Samples

An environmental replicate set generally consists of either two (duplicate) or three (triplicate) samples. The replicated samples are collected at the same field site, following the same collection procedure, and as close as possible to the same time as the original sample. The purpose of replicate samples is to document sampling and analytical precision using samples that reflect the chemistry of actual project samples. The results of analysis of sample replicates provide useful information regarding the overall ability of the field and laboratory program to quantify the constituents of interest. Differences in measured values between or among replicates reflect fine-scale temporal and spatial variability in water quality at the sample site location plus any variability or error introduced in the sample collection, sample processing, or laboratory analysis procedures. Ideally, replicates are collected and analyzed as part of the sampling protocols of every project. For some programs, replicates are collected (as backup) from every site, but only a subset of those is analyzed. Environmental samples provide a better test of precision than artificial samples because they include natural constituents that could alter the reproducibility of a given laboratory method. Precision can also be affected by bottle washing, sample collection, sample-processing procedures, and analysis.

In long-term monitoring (LTM) studies, project sites should be selected for replicate collection on a rotating basis to evaluate precision within the full variability of project samples being analyzed. For the analysis, the laboratory should alternate between analyzing a replicate set consecutively (within the same analytical run) and separating the replicate samples in analytical runs that occur on different days or at the beginning and end of the analytical run. One set of replicate project samples should be included in the sample stream at a frequency of at least 1 per 50 project samples. More commonly, this frequency should be 1 per 20 project samples if the budget allows.

## 4.3.3    Spiked Project Samples

Surface water samples tend to contain a large variety of chemical constituents with concentrations that can be highly variable. Therefore, there is the potential for one constituent to interfere with the analysis of another constituent. For example, a sample with a high concentration of dissolved organic carbon (DOC; which imparts a brown color to the water) would interfere with some analyses that rely on colorimetric measurement to determine concentrations. Well-documented methods specify which constituents may interfere with a given analysis and at what concentration range. However, these specifications should be verified for the samples within a specific project and laboratory to ensure the accuracy of the measurements. If sample concentrations are being measured in a range that could cause interference with the measurement of another constituent, the sample should be run twice, once untreated and once after being spiked with a known amount of the constituent of concern. The measured value of the sample including the spike should fall within the range of the method. If the concentration of the spiked sample equals the concentration of the unspiked sample plus the added amount, then recovery is complete, and it can be assumed that there is no interference. To express this relationship in terms of percentage recovery, the following equation can be used:

$$\% \text{ recovery} = \frac{\text{concentration}_{\text{spiked sample}} - \text{concentration}_{\text{unspiked sample}}}{\text{concentration}_{\text{added}}} \times 100 \quad (4.4)$$

The full range of an interfering constituent may not be known in the early stages of project sampling. Therefore, including spiked project samples in the sample stream is recommended until it is determined that interference is not a concern. Also, when a laboratory is starting to use a new method, the inclusion of spiked project samples is recommended to verify method specifications.

You should ask the laboratory to identify requested analyses that might be subject to interference with the types of samples that will be analyzed, intended methods, and analytes. It may not be possible, however, for laboratory staff to make that determination until after a given water body has been sampled and analyzed. If interferences are probable, then you should request analyses of spiked samples to determine the extent of interference.

## 4.3.4  External Quality Assurance Samples

Reference samples for laboratory analysis are provided by a variety of programs. These programs develop bulk samples that can be subsampled and sent to participating laboratories on a set frequency. Usually, these samples make up some type of environmental sample that is chemically similar to the samples that a laboratory typically analyzes. Results from all the laboratories are compiled, statistically summarized, and provided to the participants (usually anonymously). Participation in a reference sample program provides the opportunity for a laboratory to compare its performance with that of other laboratories. It should be included in the QA/QC program of any laboratory that analyzes environmental samples.

In particular, laboratories that analyze low-nutrient, low-ionic-strength water samples should participate in reference programs that provide these types of samples. The USGS Standard Reference Sample (SRS) Project conducts a national interlaboratory analytical evaluation program semiannually. The program includes three types of samples: low-ionic-strength, nutrient, and trace constituents. Typically, the reference samples consist of snow, rain, surface water, or DIW that is collected, filtered, and possibly spiked with reagent-grade chemicals to meet the goals of the program. Reference samples for low-ionic-strength constituents are analyzed for Ca, Cl, Mg, pH, K, Na, specific conductance, and $SO_4$. Reference samples for nutrient constituents are analyzed for $NH_4$ and $NO_3$. Reference samples for trace constituents are analyzed for Al, Ca, Mg, K, Si, and Na. Laboratory personnel are aware of the presence of the SRS at the time of analysis but do not know the constituent concentrations until results are posted on the SRS Project Web site after the conclusion of each study. The most probable value (MPV) for each constituent is equal to the median value calculated from the results submitted by participating laboratories. Laboratory results are compared with the MPV for each constituent and a percentage difference is calculated and reported.

A second standard reference program is operated by Environment Canada's NWRI. This program sends a set of 10 samples to a group of participating laboratories twice a year. The samples are obtained from predominantly low-ionic-strength waters representing several sources, such as precipitation, snow, lake water, and stream water throughout North America. The concentrations of the constituents in the NWRI samples are similar to those of the environmental samples analyzed by laboratories that specialize in low-nutrient, low-ionic-strength samples. Laboratory results are compared with an MCV calculated from results from all participants in the NWRI program. The USGS MPV and NWRI MCV are the same statistic, although named differently. Laboratory personnel are aware of the presence of NWRI samples at the time of analysis but do not know the MCV of the constituents until Environment Canada publishes a report at the conclusion of each study.

A drawback to standard reference sample programs is that the analyst knows that this is a "high-priority" sample and therefore may give extra attention to its analysis. So, the results might not fully reflect those obtained in the analysis of routine project samples. This type of analyst bias can be avoided with blind reference samples.

Blind reference samples are processed and analyzed as environmental samples and therefore appear to the analyst to be project samples. Ideally, these samples would originate from an interlaboratory reference program so that known CVs would have been or would be established for the sample. Implementation of a blind reference sample program requires the participation of one person who works in the laboratory. The reference samples must be coded and prepared by this person so that they cannot be distinguished from routine samples by the analyst. This person is also responsible for retrieving the analysis data from the laboratory database and recoding it as QA data. One blind reference sample per 50 project samples is recommended.

When evaluating candidate laboratories, their participation and performance in an interlaboratory reference program is a useful decision criterion. We recommend that, where practical, you should consider avoiding use of laboratories that do not participate in such a program. You can request and review performance results prior to making arrangements to use a particular laboratory.

## 4.4  FIELD QA

### 4.4.1  Sample Containers

The required sample containers and cleaning procedures are described in detail in the section of this book that addresses laboratory protocols. At least 2% of the cleaned containers (randomly selected) should be given a specific conductance check, which entails measuring the conductance of DIW in the sample container after a 48-h soak period. Conductance should be less than 1.2 µS/cm. If the conductance is more than 1.2 µS/cm, rerinse all the containers cleaned since the last acceptable check. If contamination is found, then 25% of the sample containers in subsequent batches should be monitored until all monitored containers in a batch pass the conductance test.

### 4.4.2  Field Measurements

Measurements of dissolved oxygen (DO), temperature, conductance, and pH are often made in the field. If these measurements are made in the field, they require field QA procedures and the use of both PE and QC samples as described in Table 4.3. These samples confirm that the measuring devices (often field meters) are functioning properly and they are in control over the entire length of the study.

Peck and Metcalf (1991) developed a stable and well-quantified (both theoretically and analytically) QC check sample for conductance, pH, and ANC measurements in dilute surface waters. It is a 1:200 dilution of the National Institute of Standards and Technology (NIST) 0.025 mol/kg $KH_2PO_4$ and 0.025 mol/kg $Na_2HPO_4$ standard pH stock solution. It has a pH of 6.89, a conductance of 37.6 µS/cm, and an ANC of 125 µeq/L. This solution is recommended as a QC check for studies doing field pH or conductance measurements in relatively well-buffered waters.

Unpublished data from the NSWS of the Environmental Protection Agency (EPA) showed that pH can be measured more precisely in the laboratory using water samples that have been collected in sealed 60-ml syringes with no more headspace than in the field using portable pH meters. In general, measurements made under controlled laboratory conditions are more precise and accurate than those made in the field, where contamination, weather, and fatigue can induce variability. Thus, we recommend that pH and conductance measurements be made in the laboratory.

Analytes that are sensitive to changes in $CO_2$ concentrations (e.g., pH, dissolved inorganic carbon [DIC]) should ideally be measured in samples collected in the field into glass bottles having septum caps or into syringes with no air bubbles and analyzed within 72 hours of collection if the sample $CO_2$ concentration is likely to be supersaturated with respect to the atmospheric

**TABLE 4.3    FIELD QUALITY CONTROL SAMPLES**

| Measurement | QC Sample Type | Description | Frequency | Acceptance Criteria | Corrective Action |
|---|---|---|---|---|---|
| Dissolved oxygen | PE sample | Concurrent determination of sample by Winkler titration | Once per meter per field season | Measured $O_2$ within ± 1 mg/L of $O_2$ estimated by Winkler titration | Replace meter or probe |
| | QC check sample | Water-saturated air | Daily (at base station) | Instrument can be calibrated to theoretical value | Replace meter or probe |
| Temperature | PE sample | Concurrent measurement of 0°C and 25°C solutions with NIST-traceable thermometer | Once per meter | Within ± 1°C of thermometer reading | Replace probe or meter |
| | QC check sample | Concurrent measurement of sample with field thermometer | Weekly | Within ± 1°C of thermometer reading | Replace probe or meter |
| Conductance | QC check sample | Solution of known conductance | Weekly | Within 10 µS/cm of theoretical value | Recalibrate meter using NIST-traceable standards; replace probe or meter |
| pH | QC check sample | Solution of known pH | Daily | Within 0.1 pH unit of theoretical value | Recalibrate meter and probe or replace probe |

*Note:* PE, performance evaluation sample.

$CO_2$ concentration. Stream samples affected by discharging groundwater (springs) and lake samples from the hypolimnion of stratified lakes are especially likely to be supersaturated with $CO_2$. Typically, you need one syringe for pH and DIC. Collecting an extra syringe is recommended in case additional sample volume is needed in the laboratory. Temperature measurements must be made in the field. DO measurements are also made in the field using a meter.

## 4.5  REPORTING QA DATA

Before selecting a laboratory, the laboratory QA results should be evaluated to ensure that the data quality delivered by that laboratory will be suitable for the planned project. The most common, and perhaps most effective, method of reporting QA data is through the use of control charts, which plot QA or QC data through time. The control charts (1) indicate whether the laboratory DQOs are met for individual QC samples; (2) reveal long-term biases or trends within and outside the control limits; and (3) provide comparisons with results from other laboratories. Each constituent has prescribed control limits that are set by the laboratory (Table 4.1). Ideally, when no bias is present, half the data points in a control chart would be above and half below the target value line. Although QC samples are used to evaluate data quality and identify samples that need to be rerun during the analysis, when plotted on control charts, QC samples also provide useful data to evaluate method performance over time, thereby also providing QA information.

Results from the analysis of QC samples are plotted on control charts in which the central line is equal to the target value (known concentration) of the control sample (Figure 4.2). Both a high- and a low-concentration QC sample, relative to the expected concentration distribution of the project samples, should be analyzed. If the QC sample is a blank, the target value is set to zero. A constituent analysis is considered biased if 70% or more of the data points on a chart are either above or below the target value line. The upper and lower control limit lines on each chart represent the range of satisfactory data based on the DQOs. The QC-high and QC-low samples are plotted on separate graphs by constituent and date of analysis, and the control charts are evaluated for trends and (or) bias and precision.

Figure 4.2 provides 3 years of data for a QC sample used for low-concentration measurements of $NO_3^-$. Virtually all of the data fall within the control limits without any indication of trends or bias. Results of the QC sample analysis shown in Figure 4.3 also indicate a reliable method; only one value fell outside the control range. However, within the control range, an upward trend occurred in 2006, followed by a downward trend in 2007. If either of these trends had continued, the data would have drifted out of control. To ensure early

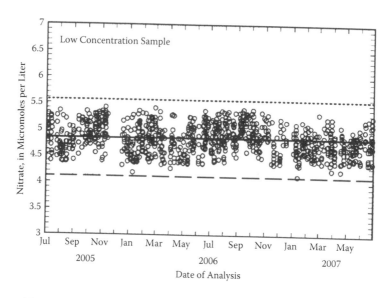

**Figure 4.2**    Results from analysis of low-concentration QC samples for nitrate analysis. The target value of the control sample is represented by the central line; the upper and lower dotted and dashed lines represent the range of satisfactory data based on the DQOs.

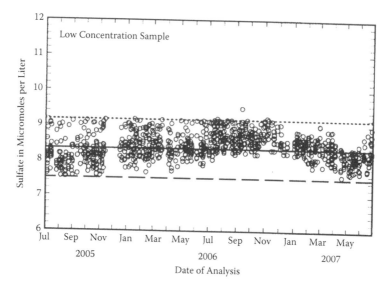

**Figure 4.3**    Results from analysis of low-concentration QC samples for sulfate analysis. The target value of the control sample is represented by the central line; the upper and lower dotted and dashed lines represent the range of satisfactory data based on the DQOs.

detection of trends in QC, control charts should be updated daily to monthly depending on the sample load. For the analysis of filter blanks and analytical blanks, the control range is defined by zero and the DQO threshold. For replicate sample concentrations, the CV of the two or three replicate samples is plotted, and the control limits are determined by plus or minus the DQO for accuracy and precision. Control charts can be used to show results from interlaboratory comparisons by plotting the percentage difference from the most-probable value. Control limits are defined by the acceptable percentage difference from the most-probable value, which might be designated to be 10%, for example, unless the concentration of the test solution is low, in which case a higher value should be selected.

Documentation of QA data should be readily available to projects that use or contemplate using the laboratory. You can request from the laboratory, in advance of starting a project, QA results that would be applicable to the planned project. These results should include, at a minimum, QC results plotted on control charts and comparisons with results from other laboratories.

## 4.6  LABORATORY AUDITS AND CERTIFICATION

All laboratories conducting chemical analyses should be periodically audited by a qualified external body on a set frequency that does not exceed 3 years. These audits should be comprehensive in covering every aspect of laboratory activities. Documentation of audit results, recommendations, and actions taken by the laboratory should be maintained and available to projects that use the laboratory. Ideally, audits are conducted as a component of a certification program. In addition, round-robin programs, in which multiple labs analyze the same set of PE samples and compare results, are an excellent way to evaluate the performance of an individual laboratory and to ensure that it provides high-quality data. If a laboratory is not able to document that it provides high-quality data, then you may want to find an alternate laboratory for analyzing samples considered to be important from a regulatory or decision-making perspective. Any laboratory chosen for project work should be able to provide documentation of audits of laboratory procedures conducted within the last 2 years.

At the time of this writing, we know of one national accreditation program for laboratories that analyze environmental samples. This program is administered through the National Environmental Laboratory Accreditation Conference (NELAC), formed on the recommendation of the EPA. NELAC is a cooperative association of state and federal agencies, formed to establish and promote mutually acceptable performance standards for the operation of environmental laboratories. The standards cover both analytical testing of environmental samples and the laboratory accreditation process. To accomplish these goals, NELAC developed the National Laboratory Accreditation Program (NELAP). NELAP recognizes state programs as accrediting authorities that administer the program. For example, a laboratory headquartered in New York State would apply to New York State's Environmental Laboratory Accreditation Program (ELAP). As each laboratory becomes accredited under a NELAP-recognized accrediting authority, the laboratory and its accredited scope of testing will be entered into a national database. One of the fundamental principles of NELAC is that of reciprocity among NELAP accrediting authorities. For example, once a laboratory is accredited by one state for testing under a specific EPA program, it can be accredited in another state for that EPA program without having to meet additional accreditation requirements. We recommend use of accredited laboratories or, at a minimum, laboratories that can demonstrate their ability to produce high-quality data, as described previously.

## 4.7  DATA ENTRY

When laboratory analyses are complete, lab results should be merged with field data and brought into the database. Import formats for lab results might include an Access database in a STORET-compatible format or an Excel spreadsheet.

Data that pass the QC checks should be electronically transferred into the laboratory database, where they can be reviewed for errors that could result from mislabeling, data entry mistakes, misidentification of samples, contamination, or a number of other potential sources of error. Part of this review process should entail examination of frequency distributions of all values determined for a given variable. Based on the frequency distribution, you can identify probable outlier values that can be rechecked for accuracy. You should perform this check for all variables. Much can be revealed by just examining spreadsheets of the data and by looking at the minimum, 25th percentile, median, 75th percentile, and maximum values of each variable. Do the numbers make sense relative to previous work in the same, or a generally similar, area? You would be amazed how often data are compromised because someone failed to perform a conversion of units or did so improperly. These kinds of simple mistakes convert your data into meaningless cybertrash. But, the good news is that such errors can be easily found and fixed. Once the data have been verified, they should be placed in a location accessible to project personnel for downloading. All types of data storage should be backed up on a daily basis.

## 4.8 SUMMARY

As discussed in the preceding sections, there are many types of QA/QC data and a variety of ways to evaluate the reliability of the data collected for a particular project. In the absence of such QA/QC analyses, it is impossible to determine whether the collected data can meet project objectives. Key elements of the QA/QC program that should be considered in evaluating a laboratory for possible use in analyzing samples from dilute lakes or streams are summarized in Table 4.4.

### TABLE 4.4   QA/QC PROTOCOL KEY ELEMENTS

1. Develop project data quality objectives (DQOs). These are project-specific goals for data quality for each measured variable. Table 4.1 is a good starting point but will not be optimal for all projects. Add or subtract any variables that are not pertinent to the project at hand. Revise criteria based on specific project needs.
   a. *Detection limits*: The detection limit is the threshold below which measured values are not considered different from zero concentration. Detection limits need to be evaluated regarding levels of ecological concern. If the detection limit is near or above the level of concern, then the usefulness of the data might be limited.
   b. *Precision and accuracy*: Precision is measured by repeated analysis of the same sample to determine the variability in the analytical data for each variable. Accuracy is measured by blind analysis of samples with known concentration. The deviation of the analytical measurements from the true known concentration is called a bias. The need for precision and accuracy for a project is dependent on the magnitude of the effect studied. If you are trying to quantify small differences among groups or small changes over time, then you will need higher precision and accuracy.
   c. *Comparability*: Data need to be comparable to what is being measured at other locations and times. To help ensure comparability, the analytical laboratory should be certified or participate in sample "round-robins" in which the same samples are analyzed by multiple laboratories.
   d. *Completeness*: Ideally, all sites intended for sampling and all measurements intended to be made at each site will be made, constituting 100% completeness. In reality, completeness is generally somewhat less than 100%. Completeness goals should be set depending on how much of a problem missing data will be in the final analysis of the data.

2. Evaluate laboratory QA data to see how well they meet your DQOs.
   a. Evaluate analytical results for blank samples.
   b. Determine results for sample replicates.
   c. Quantify expected versus observed results for spiked samples.
   d. Examine laboratory performance for external QA audit samples.

*Continued*

**TABLE 4.4 (*Continued*)    QA/QC PROTOCOL KEY ELEMENTS**

3. Evaluate field-sampling QA procedures and data.
   a. *Sample bottles*: May be provided already cleaned by the laboratory or they may need to be cleaned and tested as described in Chapter 2 on water chemistry field sampling.
   b. We recommend that water sample filtration and pH measurements be made in the laboratory, where precision is higher and risk of contamination lower, rather than in the field. If field filtration is done, field blank samples can indicate potential contamination. Water for later pH measurements in the lab should preferably be collected in 60-ml syringes without introducing any air bubbles and then sealing the syringes with a valve.
   c. Evaluate results of field QC check sample analyses for any field measurements as described in Table 4.3.

# REFERENCES

Helsel, D.R. 2005. *Nondetects and Data Analysis.* Wiley, Hoboken, NJ.
Peck, D.V. and R.C. Metcalf. 1991. Dilute, neutral pH standard of known conductivity and acid neutralizing capacity. *Analyst* 116:221–231.
US Environmental Protection Agency (EPA). 1987. *Handbook of Methods for Acid Deposition Studies: Laboratory Analyses for Surface Water Chemistry.* EPA/600/4-87/026. US EPA, Washington, DC.

# Chapter 5

## Data Analysis

### 5.1 BACKGROUND AND OBJECTIVES

After collecting your water quality samples, analyzing them in the laboratory, and quality assuring the database, it is hoped that you end up with a database that you can rely on and that accurately represents the conditions in the sampled water body at the time of sampling. Nevertheless, good data are of limited utility if they are not properly analyzed and interpreted. The data analysis is a critical part of the overall effort. Unfortunately, there is not one way to analyze your data. There are, however, certain kinds of analyses that you can conduct that are likely to shed light on the water conditions that you sampled. In this section, we highlight some of these routine analytical approaches that you might find useful. In addition to, or instead of, the analyses that we recommend, you could come up with your own variations on the themes outlined here.

Data analysis is as much an art form as a science. A good data analyst can tease a great deal of information from a database that may seem incoherent to the untrained eye. Be creative. Look at your data from multiple angles. Try to discard preconceived notions. Explore your data. Learn something new. You might be amazed by what you find. This is how science moves forward.

A data analysis protocol (DAP) is provided here. It provides the basis for translating water quality data generated in the analytical laboratory into meaningful guidance for data analysis and interpretation. It connects the raw data to the information goals for which the data were collected. There are a number of information goals that are relevant to water-sampling efforts focused on the effects of atmospherically deposited substances. These goals stem from specific questions that are often formulated to inform management decision making.

The purpose of this DAP is to describe graphical, statistical, and other approaches that can be used in validating, presenting, analyzing, and understanding water quality data. The DAP can support analytical efforts by personnel who do not have advanced training in chemistry or statistics.

However, we recommend that, whenever possible, data analysis should be conducted by individuals who have good grounding in water chemistry and a basic understanding of statistics. In particular, those involved in trends analysis should have had formal training in statistics or consult with a trained statistician before and while conducting trends analysis.

The DAP is divided into sections as follows:

1. Develop a statement of the objectives of the data analysis
2. Evaluate and ensure the quality of the dataset
3. Prepare raw data for graphical and statistical analysis
4. Conduct exploratory analyses
5. Conduct, if needed, formal statistical analyses
6. Report data in standardized formats

As described in Chapter 1 of this book, collection of surface water chemistry data should always have a purpose. Specific questions need to be formulated, and the nature of these questions will inform the design of the study, including what, where, when, and how to sample. To some degree, these questions will also help to inform how to analyze the data.

Some example approaches for data analysis that we recommend for consideration in studies of air pollution effects are outlined in Table 5.1. Each example data analysis approach given in the table is tied to a specific purpose.

## 5.2   EVALUATION OF DATA QUALITY

The first step in analyzing any raw dataset provided by an analytical laboratory should always be to conduct an evaluation of the quality of the data. This should include reviewing the quality control (QC) data provided by the laboratory, conducting quality assurance (QA) analyses, determining if data quality objectives (DQOs) have been met by the laboratory, and conducting or reviewing the results of data validation.

Data validation is the process of checking for internal consistency among data values using the ionic relationships among the analytes in the dataset. Sample contamination, analytical error, or reporting error can lead to incorrect data values that are not representative of conditions in the field. Many such errors can be identified and in some cases corrected through data validation. If data validation is done in a timely manner, problematic values can sometimes be reanalyzed in the laboratory and fixed. If nothing else, values that fail validation checks can be flagged in the dataset and considered potentially suspect in various data analyses. Sample analysis results from the laboratory are considered preliminary until the internal consistency checks described in the material that follows are performed. The database should not be released or subjected

**TABLE 5.1   EXAMPLE APPROACHES FOR DATA ANALYSIS TIED TO THE PURPOSE AND GENERAL APPROACH OF THE FIELD STUDY**

| Purpose | General Approach | Example Data Analysis |
|---|---|---|
| 1. Determine whether lake or stream is N limited | A. Measure N, P, chlorophyll *a* during snow-free season | a. Plot molar N:P ratio over time relative to published ratios that have been shown to be associated with N limitation. Compare with changes in chlorophyll *a* concentrations. |
| 2. Quantify episodic excursions from base flow chemistry | A. Measure water chemistry during rainstorms or snowmelt | a. Plot changes during episodes in discharge and selected water chemistry parameters to reveal episodic changes in ANC, pH, and Al, and to illustrate changes in potential drivers of those parameters. <br> b. Plot changes over time during spring and summer in discharge and selected water chemistry parameters. <br> c. Plot changes over time (including episodes) in the ratio of ($NO_3^-$ concentration to $[NO_3^- + SO_4^{2-}]$ concentration) versus ANC to illustrate the relative importance of $NO_3^-$ as a mineral acid anion in driving episodic ANC changes. |
| 3. Determine spatial patterns in water chemistry across a region | A. Conduct regional survey of water chemistry | a. Map various chemical concentrations across the landscape to reveal spatial patterns. Correlate those with landscape features (geology, elevation, etc.). <br> b. Examine patterns in the data across spatial gradients. <br> c. Construct histograms to identify regional outliers in concentrations of key variables. |

*Continued*

TABLE 5.1 (*Continued*)    EXAMPLE APPROACHES FOR DATA ANALYSIS TIED TO THE PURPOSE AND GENERAL APPROACH OF THE FIELD STUDY

| Purpose | General Approach | Example Data Analysis |
|---|---|---|
| 4. Quantify long-term changes over time in water chemistry | A. Sample over extended period of time (at least 8 to 10 years), preferably monthly or seasonally during open-water season | a. Standardize sampling to account for episodic changes or otherwise address variability.<br>b. Conduct trends analyses.<br>c. Compare trends in biologically relevant variables (ANC, pH, $Al_i$) with trends in potential drivers and buffers ($SO_4^{2-}$, $NO_3^-$, SBC, DOC). |
| 5. Determine extent to which air pollution is affecting water resources | A. Characterize index chemistry for multiple lakes or streams expected to be sensitive | a. Plot central tendency and variability (i.e., box and whisker plots) in key variables across sites.<br>b. Compare with common benchmark thresholds for ANC (0, 20, 50, 100 µeq/L); pH (5.0, 5.5, 6.0); and $Al_i$ (2, 7 µM) and calculate the percentage of sites that exceed thresholds. |
| | B. Conduct survey of waters in study area | a. Plot central tendency and variability in key parameters.<br>b. Compare with common benchmark thresholds. Map chemical concentrations. |

| | |
|---|---|
| C. Use dynamic model to hindcast past changes in water chemistry since about 1900 or earlier | a. Map simulated changes in chemical concentrations.<br>b. Compare simulated changes with changes derived by other means (i.e., long-term monitoring, paleolimnology, space-for-time substitution).<br>c. Plot simulated changes in ANC versus aspects of current chemistry (i.e., ANC, ANC/SO$_4^{2-}$).<br>d. Estimate proportional changes in ANC versus SO$_4^{2-}$, SBC versus SO$_4^{2-}$, ANC versus (SO$_4^{2-}$ + NO$_3^-$), and so on to reveal interactions among variables. |
| D. Collect and analyze diatom remains in sediment cores | a. Analyze as described for dynamic model hindcasts step 5C, a through c. |
| E. Simulate critical load of S or N deposition | a. Map critical loads and exceedances (amount by which ambient deposition exceeds critical load).<br>b. Plot critical load as function of water chemistry (i.e., ANC, ANC/SO$_4^{2-}$, etc.).<br>c. Develop procedures with which to extrapolate critical load on basis of water chemistry or landscape characteristics. |
| 6. Evaluate possible need for mitigation | A. Follow similar approaches as for step 5 |
| | a. Follow similar approaches as for step 5. |

*Note:*  Al, inorganic monomeric aluminum; ANC, acid-neutralizing capacity; DOC, dissolved organic carbon; N, nitrogen; NO$_3^-$, nitrate; P, phosphorus; SBC, sum of base cations; SO$_4^{2-}$, sulfate.

to data analysis and interpretation until internal consistency is evaluated and documented to the extent allowed by the laboratory analyses conducted.

Some water quality studies entail measurement of full major ion chemistry. We do not mean to infer that the only water quality database that is valid includes measurement of all major ions. Other, more limited, databases are clearly also of value. However, if measurements of all major ions have indeed been conducted, then this opens up the opportunity for a number of helpful QA checks as described in this chapter.

A variety of approaches can be used to evaluate and demonstrate overall analytical data quality. These include comparing measured with calculated variables, where each is intended to represent the same parameter. If measured and calculated values are similar, within an expected range of error, then there is an increased likelihood that the data used in the comparison are of high quality. If measured and calculated values of the same parameter differ by more than the expected variability, then it can be inferred that one or more of the values used in the calculations represented in that comparison may be in error. This approach can be helpful in flagging certain samples or measurements for reexamination to determine if there were recording errors or some other kind of error that might be identified and corrected.

Data validation protocols also include constructing plots of variables that might be expected to correlate with each other. Any sample that deviates substantially from the expected relationship might be further examined for potential error or flagged in the dataset. A flag placed on a data value in a dataset signifies to the user that there may be decreased confidence in that particular value. The data analyst can then use his or her discretion in deciding how to deal with that decreased confidence.

## 5.2.1  Charge Balance

The sum of positively charged ions (cations) in water must equal the sum of those with negative charge (anions). Otherwise, the water might spontaneously burst into flames (just kidding!). This is called the principle of electroneutrality. Major discrepancies between the sum of measured anions and cations thus reflect analytical errors, failure to measure all ions with significant concentrations, or a combination of both. Although charge balance calculation and comparison alone cannot necessarily identify the cause of a charge imbalance, they can serve as a QA check on the completeness and accuracy of the ion chemistry data. A high-quality dataset will show reasonable agreement between the calculated cation and anion sums, after accounting for the failure to measure all ions in solution, in particular organic acid anions.

To assess the quality of the data for the ionic species in water, ion charge balances involving a comparison of the sum of cations to the sum of anions should be calculated for all water samples subjected to full ion measurement. If the data are provided in mass units (e.g., mg/L), then they must be first converted into equivalence (i.e., microequivalents per liter, µeq/L) units by multiplying the concentration in milligrams per liter by the appropriate factor in Table 5.2. Anion and cation sums in units of milliequivalents per liter are then approximated as defined in Equations 5.1 and 5.2:

$$\text{Sum of cations} = Ca^{2+} + Mg^{2+} + Na^+ + K^+ + NH_4^+ + H^+, \tag{5.1}$$

$$\text{Sum of anions} = SO_4^{2-} + NO_3^- + Cl^- + F^- + (ANC + H^+). \tag{5.2}$$

**TABLE 5.2   FACTORS FOR CONVERTING MILLIGRAM-PER-LITER (mg/L) UNITS OR pH UNITS TO MICROEQUIVALENT-PER-LITER (µeq/L) UNITS**

| Analyte | To Convert from mg/L to µeq/L, Multiply by |
|---|---|
| $Ca^{2+}$ | 49.90 |
| $Mg^{2+}$ | 82.29 |
| $Na^+$ | 43.50 |
| $K^+$ | 25.58 |
| $NH_4^+$ (mg $NH_4$/L) | 55.44 |
| $NH_4^+$ (mg N/L) | 71.39 |
| $SO_4^{2-}$ | 20.82 |
| $SO_4^{2-}$ (mg S/L) | 62.38 |
| ANC (mg $CaCO_3$/L) | 19.98 |
| $Cl^-$ | 28.21 |
| $F^-$ | 52.63 |
| $NO_3^-$ (mg $NO_3$/L) | 16.13 |
| $NO_3^-$ (mg N/L) | 71.39 |
| $Al^{n+}$ (inorganic monomeric) | 74.13[a] (assumes a +2 charge) |
| $DOC$[b] | 5 to 10 (rough approximation) |

*Note:*  To convert pH to $H^+$ in units of µeq/L: $H^+$ (µeq/L) = $10^{-pH} \times 1,000,000$.

[a] The factor given for conversion of Al concentration from mass units to equivalence units assumes an average charge of +2 on the inorganic Al species present in the water. If the water sample has very low pH (less than about 4.8), then use a factor of 111.19 instead of 74.13 (assumes average charge of +3).

[b] To convert DOC in mg/L to DOC concentration in µmol/L, multiply by 83.33. To estimate the equivalent concentration of organic acid anions from the DOC, multiply DOC in mg/L by a value of 5 to 10 to generate a very rough estimate of organic acid anion concentration in µeq/L.

The (ANC + H⁺) term in the anion sum is determined by laboratory measurements of acid-neutralizing capacity (ANC) and pH. The hydrogen ion concentration ($H^+$) is calculated from pH as shown in Table 5.2. To make this conversion on a calculator, take the pH value, change the sign to negative, hit the inverse $\log_{10}$ key, and multiply by $10^6$.

A charge balance plot should be made by plotting the sum of anions versus the sum of cations (Figure 5.1). Some deviation from the one-to-one line (i.e., $y = x$) is expected because of analytical errors associated with the measurement of the individual anions and cations. Although random analytical errors would tend to cancel in calculating the sum of anions or cations, the analytical accuracy and precision and their relative contribution to the ion sum differ for each of the ions measured. Hence, some charge imbalance may occur because of differences in analytical precision and accuracy. Thus, the calculated charge balance is an imprecise measure of data quality. It is useful as a tool for determining rather large deviations from the expected relative concentrations of anions and cations.

Percentage ion balance difference [% IBD; (Cation sum – Anion sum)/(Cation sum + Anion sum) × 100] should be calculated for all samples. As a general guideline, we recommend the following criteria for % IBD (Table 5.3):

If the sum of anions + cations ≤ 100 µeq/L, % IBD should be ≤ 25%.

If the sum of anions + cations > 100 µeq/L, % IBD should be ≤ 10%.

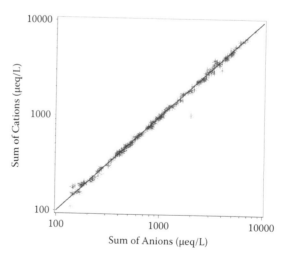

**Figure 5.1**    Example plot of cation sum versus anion sum. There are two potential outliers that warrant further investigation to determine if an error was made in analyzing or reporting the concentration of one or more ions for these two samples.

**TABLE 5.3    DATA VALIDATION QUALITY CONTROL PROCEDURES**

| Activity or Procedure | Requirements and Corrective Action |
|---|---|
| Range checks, summary statistics, frequency distributions, or other exploratory data analysis (e.g., box and whisker plots). | Identify suspect values. Review field notes and laboratory data for possible problems or errors. Correct reporting errors or qualify as suspect or potentially invalid. |
| Ion balance: Calculate percentage ion balance difference (% IBD) using data from cations, anions, pH, and ANC. | If total ionic strength[a] $\leq$ 100 µeq/L, % IBD should be $\leq$ 25%. If total ionic strength > 100 µeq/L, % IBD should be $\leq$ 10%. Determine, if possible, which analytes are the largest contributors to the ion imbalance. Review suspect analytes for possible analytical error and reanalyze any samples for which analytical error appears likely. If analytical error is not indicated, qualify sample to attribute imbalance to unmeasured ions. Reanalysis is not required. |
| Conductivity check: Compare measured conductivity of each sample to a calculated conductivity based on the equivalent conductances of all major ions in solution. | If measured conductivity $\leq$ 25 µS/cm, ([Measured – Calculated] ÷ Measured) should be $\leq$ ±25%. If measured conductivity > 25 µS/cm, ([Measured – Calculated] ÷ Measured) should be $\leq$ ±15%. Determine, if possible, which analytes are the largest contributors to the difference between calculated and measured conductivity. Review suspect analytes for analytical error and reanalyze any samples for which analytical error appears likely. If analytical error is not indicated, qualify sample to attribute conductivity difference to unmeasured ions. Reanalysis is not required. |
| Aluminum check: Compare results for organic monomeric aluminum and total monomeric aluminum. | [Organic monomeric] should be < [total monomeric]. Review suspect measurements to confirm if analytical error is responsible for inconsistency. |

*Continued*

**TABLE 5.3 (*Continued*)   DATA VALIDATION QUALITY CONTROL PROCEDURES**

| Activity or Procedure | Requirements and Corrective Action |
|---|---|
| **ANC check** | |
| 1. Calculate carbonate alkalinity based on pH and dissolved inorganic carbon (DIC). Compare to measured ANC.<br>2. Calculate charge balance ANC and compare with laboratory-measured (titrated) ANC. | Review suspect measurements for samples with titrated ANC < carbonate alkalinity or those with differences > 15% or > 15 μeq/L for samples with ANC < 150 μeq/L. Determine if data entry error, analytical error, or noncarbonate alkalinity is likely to be responsible for lack of agreement.<br>Review samples having ([Measured – Calculated] ÷ Measured) > ± 15% (for low-DOC waters) to 20% (high-DOC waters). Determine if observed discrepancy can be attributed to organic anions.<br>Strong organic acid anions are expected to decrease titrated ANC, compared with calculated charge balance ANC, by an amount equal to approximately (as a crude approximation) 5 times the DOC concentration in mg/L for acidic waters (ANC < 0 μeq/L) to 10 times the DOC concentration in mg/L. Determine, if possible, if data entry error or analytical error is likely to be responsible for the observed inconsistency. |

*Source:* Modified from Paulsen, S. 1997. *Environmental Monitoring and Assessment Program: Integrated Quality Assurance Project Plan for Surface Waters Research Activities.* US Environmental Protection Agency, Office of Research and Development, National Health and Environmental Effects Research Laboratory, Corvallis, OR.

[a] Total ionic strength is calculated as the sum of cations (Equation 5.1) added to the sum of anions (Equation 5.2).

For any samples that do not satisfy these criteria, the analytical data should be reviewed to determine if the cause of the imbalance is data entry error, analysis error, or some other identifiable error. If laboratory analysis error is discovered, the sample should be reanalyzed for those analytes that do not exceed laboratory holding times. If no error can be determined through data review or reanalysis, the results are finalized without change, assuming that the imbalance is caused by unmeasured ions.

Although the calculated charge balances do not include all ions that could potentially contribute to the sum of the cations and anions, those that are included contribute most to the overall anion and cation sum in most dilute freshwater environments. Inorganic ions not included, such as phosphorus (P) and trace metals, are generally present in relatively low concentrations in most low-ionic-strength waters and are not significant contributors to the total ion balance. Silica is not included in the charge balance because in most natural waters it exists predominantly in an uncharged form and does not contribute to either the anion or cation charge balance.

Two types of water bodies, however, often have charge imbalance caused by ions that are not included in Equations 5.1 and 5.2. In acidic waters (pH less than about 5.5), aluminum (Al), which becomes more soluble with decreasing pH, may be a major contributor to the cation sum. Also, in waters with relatively large amounts of dissolved organic carbon (DOC; higher than about 3 mg/L [250 µmol/L] to 5 mg/L [417 µmol/L]), organic anions can be a major contributor to the anion sum.

In acidic waters, failure to include Al in the charge balance may cause a cation deficit (anions greater than cations). At pH greater than 5.5, Al concentrations are typically so low that they are unimportant in the overall ion balance. At lower pH, however, Al should be incorporated into the cation sum for charge balance checks. Typically, there are several different forms (species) of inorganic Al, and they can have different charges. The concentration of Al (µeq/L) can be approximated by converting measured values of inorganic monomeric Al ($Al_i$) in milligrams per liter to equivalence units using the conversion factor in Table 5.2, which assumes a +2 average charge for Al. For highly acidic waters (pH less than about 4.8), an average charge of +3 should be assumed for Al (as given in the footnote to Table 5.2). Alternatively, if Al is not measured, a cation deficit (anion sum higher than cation sum) in acidic waters (pH less than about 5.2) should not necessarily be interpreted as a QA problem because it can be assumed that some or all of the cation deficit results from unmeasured Al.

Naturally occurring organic anions (derived from organic acids) contribute to the overall anion sum. Because there is no direct measure of organic anions, they are typically not included in the anion sum as represented in Equation 5.2. Where they are present in significant concentrations, the charge balance will show an anion deficit (cation sum higher than anion sum). DOC concentration may be used as a surrogate variable for organic anions to check whether any observed anion deficit could be related to organic acids. In general, when DOC is less than about 3 mg/L (250 µmol/L) to 5 mg/L (417 µmol/L), organic anion contributions to the ion balance are relatively minor and can be ignored. When DOC is greater than about 5 mg/L, there should be an appreciable anion deficit, calculated with the following equation:

$$\text{Anion deficit} = (\text{Cation sum}) - (\text{Anion sum}) \qquad (5.3)$$

The anion deficit should be roughly proportional to the DOC, with higher anion deficit in samples having higher DOC, and to some extent also higher pH. In general, the slope of the plot of anion deficit (in $\mu eq/L$; $y$ axis) versus DOC (in $mg/L$; $x$ axis) should be about 5 to 10 $\mu eq$ of anion deficit per milligram of DOC.

## 5.2.2   Calculated versus Measured Conductivity

The presence of ions in water increases the electrical conductivity (also called specific conductance) of that solution. Conductivity, therefore, provides an indication of total ion concentration. Further, because the relationship between ion concentration and conductivity is known for most ionic species, the measured conductivity of a water sample can be used as an internal check on both the accuracy and the completeness of the measurements of ionic species by comparing the measured and expected conductivity. The expected conductivity is calculated as the sum of the product of the ionic concentration times the equivalent conductances of each of the measured ions in water. For relatively dilute waters (conductivity below 200 $\mu S/cm$), Equation 5.4 is used. For higher-conductivity waters, a more complex equation is used, which adjusts for high-concentration effects. All waters that are sensitive to acidification from acidic deposition and most waters that are sensitive to nutrient enrichment effects from atmospheric nitrogen (N) deposition will have conductivity less than 200 $\mu S/cm$. Thus, we recommend use of Equation 5.4 and do not present the more complex equation. All of the concentrations in the equation need to be in units of microequivalents per liter. Conversion factors to convert from mass units to equivalence units are given in Table 5.2. For samples having conductivity less than 200 $\mu S/cm$,

$$
\begin{aligned}
\text{Calculated conductivity} = (&(Ca^{2+} \times 59.47) + (Mg^{2+} \times 53.0) + (K^+ \times 73.48) \\
&+ (Na^+ \times 50.08) + (NH_4^+ \times 73.5) + (H^+ \times 349.65) \\
&+ (SO_4^{2-} \times 80.0) + ((ANC + H^+ - OH^-) \times 44.5) \\
&+ (Cl^- \times 76.31) + (NO_3^- \times 71.42) + (OH^- \times 198))/1000
\end{aligned}
$$

$$(5.4)$$

Calculated conductivity should be plotted against measured conductivity as a first step to look for gross outliers (data values that fall well outside the normal range; Figure 5.2). As a more quantitative QA check, the percentage conductivity difference should be calculated as

$$\text{\% Conductivity difference} = (\text{Calculated} - \text{Measured})/\text{Measured} \times 100 \quad (5.5)$$

As a general guideline, we recommend careful review of samples for which the percentage conductivity difference exceeds 25% for samples in which measured conductivity is less than 25 $\mu S/cm$. We further recommend careful review of samples for potential data entry or analysis error if the percentage conductivity difference exceeds 15% for samples in which measured conductivity is more than 25 $\mu S/cm$ (Table 5.3).

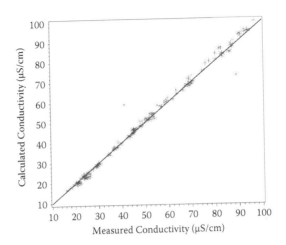

**Figure 5.2**   Example plot to examine the relationship between calculated and measured conductivity. There are two obvious outliers that warrant further investigation to determine if an error was made in analyzing or reporting the concentration of one or more ions for those two samples.

## 5.2.3   Calculated versus Measured ANC

### 5.2.3.1   Carbonate Alkalinity versus Titrated ANC

There are two methods for evaluating the internal consistency of the dataset on the basis of observed differences between laboratory measurements (titrations) of ANC and calculated ANC or carbonate alkalinity using various ion measurements. The first involves comparisons between calculated carbonate alkalinity and laboratory measures of ANC made by acid titration. In almost all surface waters, the vast majority of ANC is made up of carbonate alkalinity (Figure 5.3). Carbonate alkalinity ($[Alk_c]$) is calculated directly from laboratory measurements of pH and dissolved inorganic carbon (DIC) concentrations (Hillman et al. 1987) and is a measure of just the carbonate ions ($HCO_3^-$ and $CO_3^{2-}$) in the sample that would react with acid during an ANC titration. $Alk_c$ (in $\mu eq/L$) is calculated from the equation

$$Alk_c = ((DIC/12011) \times ((Hmolar \times K_1 + 2 \times K_1 \times K_2)/(Hmolar \times Hmolar + Hmolar \times K_1 + K_1 \times K_2)))((K_W/Hmolar) - Hmolar) \times 10^6, \qquad (5.6)$$

where DIC is in milligrams per liter, Hmolar $= 10^{-pH}$, $K_1 = 4.4463 \times 10^{-7}$, $K_2 = 4.6881 \times 10^{-11}$, and $K_W = 1.01 \times 10^{-14}$.

ANC is a measure of all ions that react with acid during the acid titration. It includes all the carbonate ions that are represented in $Alk_c$. Thus, ANC must be greater than or equal to $Alk_c$. If calculated $Alk_c$ is higher than the titrated ANC,

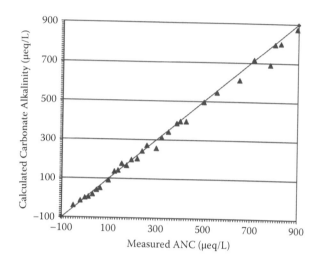

**Figure 5.3** Example plot of calculated carbonate alkalinity versus laboratory-titrated ANC. In general, calculated carbonate alkalinity is slightly lower than measured ANC (as it should be). There are no obvious outliers in this hypothetical example.

the discrepancy must be caused by analytical errors in the measurement of ANC, pH, or DIC; the presence of noncarbonate ions that react with acid during the titration; or a combination of both. Comparisons of the $Alk_c$ and titrated ANC can serve as a QC check on the measured pH, DIC, and ANC. In nonacidic (ANC greater than 0 µeq/L) waters with low DOC, samples that have ANC less than $Alk_c$ or those with (ANC – $Alk_c$) differences greater than 15% (greater than 15 µeq/L for samples with ANC less than 150 µeq/L) should be carefully reviewed for potential QA problems (Table 5.3). Acidic waters (ANC less than 0 µeq/L) and higher-DOC waters (above about 3 mg/L [250 µmol/L] to 5 mg/L [417 µmol/L]) often have other ions (Al, weak organic acid anions) that react with acid during the titration. Therefore, [ANC–$Alk_c$] differences do not necessarily indicate a QA problem, but they do suggest that the data should be reviewed for potential errors.

## 5.2.3.2 Calculated versus Titrated ANC

The second QA check of ANC values compares laboratory Gran-titrated ANC ($ANC_G$) with a charge balance definition of ANC ($ANC_{calk}$), calculated as

$$ANC_{calk} = Ca^{2+} + Mg^{2+} + K^+ + Na^+ + NH_4^+ + Al^{2+} - SO_4^{2-} - NO_3^- - Cl^-  \quad (5.7)$$

where all parameters are expressed in units of microequivalents per liter. For Al species, assume an average charge of +2 (thus, Al concentration in microequivalents per liter equals Al concentrations in micromoles per liter times 2) for waters having pH above about 4.8. For waters having lower pH, assume an average charge on the Al species of +3 (thus, Al concentration in microequivalents per liter equals Al concentration in micromoles per liter times 3). For low-DOC (less than about 3 mg/L [250 μmol/L] to 5 mg/L [417 μmol/L]) waters, laboratory-titrated and -calculated charge balance ANC should be approximately equal, plus or minus an allowance for analytical errors. In general, the errors on the individual ions should cancel each other out, and the two estimates of ANC should be within about 15% of each other (or within about 15 μeq/L for relatively low-ANC [less than 50 μeq/L] waters). If they differ by more than this amount, it suggests errors in one or more of the measurements used in the calculations and the comparison.

For higher-DOC waters (greater than about 3 mg/L [250 μmol/L] to 5 mg/L [417 μmol/L]), laboratory-titrated ANC should be lower than calculated charge balance ANC by an amount equal to the concentration of strong organic acid anions in solution. That concentration of strong organic acid anions can be roughly approximated by multiplying the DOC concentration (expressed in milligrams per liter) by the estimated organic acid charge density (average charge per milligram of DOC). Thus, for high-DOC waters, DOC-adjusted titrated ANC is calculated as

$$\text{DOC-adjusted ANC}_G = Ca^{2+} + Mg^{2+} + Na^+ + K^+ + NH_4^+ - SO_4^{2-} - NO_3^- - Cl^- - A^-$$
$$(5.8)$$

where $A^-$ is the estimated strong organic acid anion concentration (defined as those with acid dissociation constants giving them an equilibrium pH less than about 4), which is very roughly approximated by

$$A^- \ (\mu eq/L) \approx DOC \ (mg/L) \times 4 \ \mu eq/mg \qquad (5.9)$$

If the DOC-adjusted laboratory-titrated ANC differs from calculated charge balance ANC by more than about 20% (or 20 μeq/L for relatively low-ANC [≤ 50 μeq/L] samples) in high-DOC waters, that suggests the possibility of data entry error or analytical error in one or more of the parameters that enter into the calculations. In that case, laboratory and data entry records should be reviewed for possible errors.

Note that these methods for estimation of the equivalent concentrations of Al and strong organic acid anions are crude approximations for the purpose of evaluating the internal consistency of the dataset and for identifying possible incorrect values for further examination. For high-DOC waters, in particular, lack of agreement between calculated and titrated ANC does not necessarily mean that there are errors in the dataset. More rigorous approaches are available for calculating the equivalent concentrations of Al and organic anions, but these are not needed for the purpose of dataset validation.

### 5.2.4 Other Validation Procedures

The dataset should be examined in other ways to look for outliers (data values that fall well outside the normal range for that water body over multiple samplings or for multiple water bodies within a region). The range of values in the dataset or a histogram plot (Figure 5.4) should be used to look for outliers in all variables. Outliers may also be identified by plotting each variable by sample date to look for isolated gross variations over time. Analysis and sample collection records should then be reviewed to determine if the cause of any outlier is likely data entry error, analytical error in the laboratory, or sampling error in the field. If errors are discovered, samples can be reanalyzed or rejected. If no error can be determined, the results should be assumed to be correct and accepted without change. Outliers should not be rejected unless there is a strictly objective basis for rejection. If there is a clearly identified error, the result should be rejected and if possible corrected; if there is an unexplained anomaly, the data should be retained.

Another useful procedure is to plot variables in the dataset against each other for variables that are known to be highly correlated. Examples of strongly correlated variables can include $Ca^{2+}$–$Mg^{2+}$, $Ca^{2+}$–ANC, $Na^+$–$Cl^-$, DOC–color, and N–P. Data points that fall outside the cloud of data points defining the general relationship warrant closer examination.

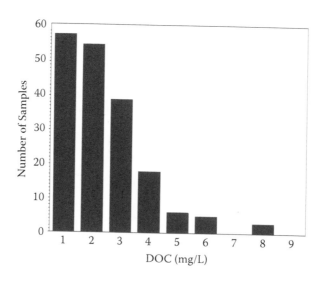

**Figure 5.4** Example histogram plot of patterns in DOC concentration in streams within a particular region. The data are not normally distributed; rather, they are skewed toward lower concentrations ($\leq 3$ mg/L).

## 5.2.5  Final Data Quality Determination

Each of the dataset internal consistency checks outlined provides an opportunity to identify potential problems in the data related to data entry error or laboratory error. Some of the problems identified through these analyses might be corrected by reanalysis or simply by replacing a value that was entered incorrectly into the dataset. In other cases, the cause of the anomaly will be unknown and will represent an error of some sort or the presence of one or more unmeasured analytes. Such unexplained deviations from expected patterns should generally not be altered or deleted from the dataset. As described in the preceding material, the final validated dataset should show

1. Good charge balance agreement
2. Good agreement between measured and calculated conductivity
3. Good agreement between laboratory-titrated and defined ANC based on various ion measurements
4. Reasonable (readily understandable) distribution of parameter values as reflected in frequency distributions across space or across time
5. Clear patterns between paired variables that are known to be strongly correlated with each other

Note that there is generally no clear-cut definition of what constitutes "good" or "reasonable" agreement, although targets for percentage and absolute variation are presented where applicable. The purpose of these analyses is not to discard measurements that are not completely understandable but rather to identify the samples or measurements that appear to have a higher likelihood of some kind of error. In the best of cases, the error is identified and corrected. In other cases, the error remains unknown, or there may not be an error at all, but rather an aspect of the water chemistry that is not fully understood.

Unless these internal consistency checks are conducted and unless the dataset is found to be generally internally consistent, it is not possible to determine whether analysis of these data will yield meaningful and representative results. This is a critically important, and frequently overlooked, aspect of water quality study. To the extent that the water chemistry data make sense, greater confidence can be placed in conclusions drawn from analysis of those data.

## 5.3  APPLY PROCEDURES TO PREPARE RAW DATA FOR GRAPHICAL AND STATISTICAL ANALYSIS

### 5.3.1  Censored Data

Data that have reduced certainty are often censored for reporting or analysis purposes or both. Examples can include measured values below the method detection limit (MDL) or measured values below the reporting limit (e.g., if nitrate concentration is reported as less than 1 µeq/L). Censored data can cause problems in statistical calculations if there is no real number that can be used in the calculations. Other problems can arise in deciding what to do with censored data when reporting limits are approximately the same as analyte concentrations of ecological concern. If detection and reporting limits are well below any real level of concern, there is generally no substantial problem with interpretation or treatment of censored data. However, when detection or reporting limits are approximately at or below the same level as the level of concern, then interpretation based on censored data may be problematic regardless of how censored data are handled.

Various recommendations can be made concerning database reporting or censoring of laboratory measurements below MDL and reporting limit values. Regardless of how a particular project decides to handle such measurements, reporting as zero, flagging data, or reporting the obtained measurement result, we recommend that MDL values and any calculated reporting limits associated with analysis results should be included with the database or otherwise made available to data users.

## 5.3.2  Outliers and Missing Values

Outliers can be difficult to identify and interpret. Their importance is dependent on the type and objective of the analysis being conducted. There are a number of statistical outlier tests that one could apply, but we caution against removing any outlier unless there is a good argument for removal based on a clearly identified analytical error or data management issue (e.g., typographical error). It is often useful to run the statistical analysis with and without any suspected outliers to see if the results are substantially different. If not, then the point is moot, and it does not matter. If the outlier causes a big change in the results, then it will require some careful reexamination of the data before deciding what to do. In general, any conclusion that relies on the presence of one or a few extreme values should not be considered a robust conclusion. If it is determined that an important conclusion does depend on the inclusion of one or a few data outliers, we recommend *not* drawing that conclusion but rather going back into the field to collect additional data.

In general, because outliers are *not* routinely removed from the dataset, outliers can be subjectively identified visually, without a formal statistical test. However, if one or more measured values appears to be an outlier, the analyst may wish to eliminate this deviant value and not include it in various calculations, data analyses, or data presentations. Results should have a caveat to reveal that one or more outliers were deleted from that particular analysis. This generally should not be done unless it can be objectively determined that the questionable value is indeed likely to be erroneous. We recommend Dixon's $Q$ test as a relatively simple test to determine outlier status. The test is conducted as follows:

1. Arrange the values of the observations in ascending order.
2. Calculate the experimental $Q$ statistic $Q_{exp}$ as the ratio of the difference between the suspect value and the value of its nearest neighbor (in the ascending series of values that make up the dataset) divided by the range of values in the dataset. For example, to test whether the lowest value $x_1$ is an outlier, calculate $Q_{exp}$ as

$$Q_{exp} = \frac{x_2 - x_1}{x_N - x_1}. \tag{5.10}$$

where $x_1$ is the lowest value in the series, $x_2$ is the second-lowest value in the series, and $x_N$ is the highest value in the series
    Similarly, to test whether the highest value is an outlier,

$$Q_{exp} = \frac{x_N - x_{N-1}}{x_N - x_1}. \tag{5.11}$$

where $x_{N-1}$ is the second-highest value in the series, and the other terms are as defined previously.
3. Compare the calculated $Q_{exp}$ to a critical $Q$ value $Q_{crit}$ that is taken from a table (Rorabacher 1991; see, for example, Table 5.4). You must first choose your confidence level (CL). We recommend a 95% CL. For example, as shown in the table, at the 95% CL and a total number of measured values equal to 9, the $Q_{crit}$ is 0.493.
4. If $Q_{exp} > Q_{crit}$, the questionable value can be designated as an outlier.

Missing values are a fact of life in most statistical analyses of environmental data. They are more problematic in parametric tests (regression, analysis of variance [ANOVA]) than in nonparametric tests. Parametric tests are used for estimating parameter values and testing hypotheses concerning them when the form of the underlying data distribution is known (typically, the data are normally distributed). For tests of data for which we do not know the underlying data distribution, including those that are not normally distributed, nonparametric tests must be used. These tests compare the distributions, rather than the parameters.

Missing values can be synthesized from other data, but we would not recommend this approach as a general procedure without careful consideration. The National Surface Water Surveys of the Environmental Protection Agency (EPA) synthesized a small number of missing values using regional regression models to make complete regional population estimates. The general approach is to substitute for the missing value a synthetic value developed from the remainder of the dataset or published relationships. The synthetic value can be

**TABLE 5.4    CRITICAL Q VALUES FOR DIXON'S OUTLIER Q TEST AT THE 0.95 CONFIDENCE LEVEL**

| Number of Measurements | $Q_{crit}$ (CL: 95%) | Number of Measurements | $Q_{crit}$ (CL: 95%) |
|---|---|---|---|
| 3 | 0.970 | 17 | 0.365 |
| 4 | 0.829 | 18 | 0.356 |
| 5 | 0.710 | 19 | 0.349 |
| 6 | 0.625 | 20 | 0.342 |
| 7 | 0.568 | 21 | 0.337 |
| 8 | 0.526 | 22 | 0.331 |
| 9 | 0.493 | 23 | 0.326 |
| 10 | 0.466 | 24 | 0.321 |
| 11 | 0.444 | 25 | 0.317 |
| 12 | 0.426 | 26 | 0.312 |
| 13 | 0.410 | 27 | 0.308 |
| 14 | 0.396 | 28 | 0.305 |
| 15 | 0.384 | 29 | 0.301 |
| 16 | 0.374 | 30 | 0.298 |

*Source:* Reprinted, with permission, from Rorabacher, D.B. 1991. *Anal. Chem.* 63(2):139–146. Copyright 1991 American Chemical Society.

calculated as the median of the existing measured values for that parameter or using a regression relationship based on one or more other variables. For example, if a measured value of $Na^+$ is missing, one can estimate the missing $Na^+$ concentration from the measured $Cl^-$ concentration using a regression approach based on $Na^+$ and $Cl^-$ measurements in the dataset. Thus, the regression equation, developed from the existing data, with which to estimate the $Na^+$ concentration from the measured $Cl^-$ concentration, should be used to estimate any missing values of $Na^+$ concentration. Similarly, inorganic monomeric Al concentration can be estimated from pH or $H^+$ concentration using a linear regression approach.

The median of existing measured values can also provide a reasonable substitute for a missing value. However, one should be careful to avoid using the median of data points known to exhibit a wide range of values, especially when there is an opportunity to reduce that variability. For example, if a $Ca^{2+}$ concentration measurement is missing from a dataset containing first- through fifth-order streams and where the $Ca^{2+}$ concentration varies strongly with stream order, it would be better to take the median of all streams in the dataset that are of the same order as the stream having the missing value rather than the median of all streams of all stream orders.

It can be considered acceptable to create synthetic substitutes for a small number of missing values, but these should generally not constitute more than 5% of the data for any variable. In general, we recommend not creating synthetic substitutes for missing values unless these missing values prevent the use of a particular analysis needed for a project objective. For example, a principal components analysis (PCA) cannot be performed on lake ion chemistry using samples that have one or more missing variable values. Thus, any sample that has even one missing value cannot enter into the analysis unless the void is first filled with a synthetic value.

Some missing values may not be particularly important to interpretation of the data (for example, a missing $NH_4^+$ concentration in a lake, which is expected to be very low). It is advisable to avoid, if possible, the need to delete that entire sample from the analysis simply because the $NH_4^+$ measurement is missing. If synthetic values are to be constructed, we recommend using whatever is the most robust empirical approach that can be developed from that particular dataset.

### 5.3.3  Multiple Observations

We do not recommend averaging the results of replicate (duplicate or triplicate) samples in the dataset. Rather, the first sample collected at a given site and sample occasion is considered to be the normal sample. It is used in statistical and other data analyses to represent the chemistry of that lake or stream on that sampling occasion. Any second or third sample (replicate) collected on that sampling occasion is used only for QA purposes, to assist in quantifying the cumulative variability and error associated with the collection and laboratory analysis of the water in that lake or stream. The replicate sample results are not used in routine data analyses.

If multiple samples are collected within a given rainstorm, season, or year, results of analyses of those samples are maintained as separate values in the dataset. Depending on the objectives of a particular study or analysis, they might be averaged in the process of analyzing the data. For example, if the objective is to compare spring base flow chemistry across streams in a particular study area, one may choose to average all samples collected during the spring season (avoiding rainstorm and snowmelt periods) over a finite period of time (perhaps 5 years). Such an approach is appropriate if, for example, some streams were sampled only once and others were sampled multiple times within that 5-year period. If there is reason to believe that stream chemistry changed appreciably during that 5-year period, then it may not be advisable to average the data across multiple years. Instead, one may choose to use the spring base flow sample collected at the time closest to April of a particular year, for example. One should be particularly careful about averaging multi-year data if part of the sampling window occurred during unusually wet years and part during unusually dry years.

If most or all sample sites were sampled during multiple years, an analysis of the spatial distribution of water chemistry across the study area will often be conducted using 3- or 5-year averages of chemistry to represent each site. Such averages should not combine samples collected during different seasons unless it is clear that seasonality is not an important issue.

### 5.3.4  Treatment of Zeros and Negative Values

For studies of dilute surface waters potentially impacted by air pollutants, the only major variable expected to on occasion have negative values is ANC. Some, but not all, lake or stream datasets will have some negative ANC values. Because negative ANC values are real measurements, they must be left as negative numbers. However, some transformations (e.g., log transforms) required for some statistical analyses may only be applicable to nonnegative and nonzero numbers. If that becomes an issue for an analysis planned for a particular dataset, add a constant whole number just larger than the largest negative number in the data (i.e., add $50\,\mu eq/L$ if the lowest ANC is $-49\,\mu eq/L$) to all ANC measurements so that there are no longer any negative numbers in the analysis. This should be done only for that particular analysis in question. Designate the new variable as [ANC + 50 $\mu eq/L$]. This manipulation of the data must be taken into consideration in interpreting the results of the analysis. For zero values, we recommend adding 1 to all values of that variable when almost all the data are greater than 1 and changing the name of the revised variables to be used in the analysis to, for example, (sodium + 1 $\mu eq/L$). This works well for $\log_{10}$ transforms as when $x = 0$, $\log_{10}(x+1) = 0$. When many of the data values are less than 1, add a constant number that is smaller than almost all the data values to each zero value in the dataset. For example, zero values for $NH_4^+$ concentration, which may be fairly common in many surface water datasets, may be adjusted by adding a constant of 0.001 $\mu eq/L$ prior to transformation.

### 5.3.5  Treatment of Seasonality

Seasonal variation in water chemistry data can affect data analysis and interpretation in two fundamental ways. First, chemical parameters that affect the suitability of the water to support various species and biological communities tend to vary with season. This is the case in many waters with respect to pH, ANC, $Al_i$, DOC, $NO_3^-$, $SO_4^{2-}$, and base cation concentrations. Thus, the chemical conditions that are most stressful to biota may occur to a greater or lesser degree depending on season. These seasonal differences are most pronounced in regions that experience substantial seasonal changes in rainfall or temperature.

Interpretation of chemical parameter values above or below known biological stress thresholds will be highly influenced by when the samples were collected.

Second, seasonality in the data can affect certain statistical tests, such as trends analysis, for example. A dataset having substantial seasonality may require use of different statistical tests as compared with a dataset lacking seasonality. This is further discussed in Section 5.5 of this DAP.

This DAP does not recommend the need for any particular adjustment of seasonal data. Some sampling studies may choose to reduce the effects of seasonality on the data by careful timing of field activities. Other studies may strive to quantify the seasonality that occurs. It can also be useful to quantify the relationship between annual average or base flow chemistry and observed extreme values that are influenced by season or episodic processes. For example, Sullivan et al. (2003) illustrated the relationship between median spring season ANC and the minimum ANC reported in the data record for streams in Shenandoah National Park (Figure 5.5). Such an analysis could also be conducted to compare spring median or spring minimum ANC with summer or

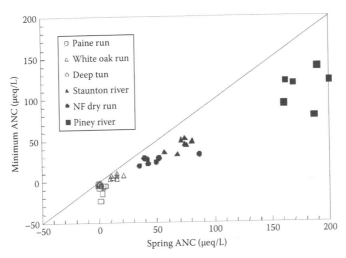

**Figure 5.5**  Minimum stream ANC sampled at each site during each year versus median spring ANC for all samples collected at that site during that spring season. Data are provided for all intensively studied streams within Shenandoah National Park during the period 1993–1999. A 1:1 line is provided for reference. The vertical distance from each sample point upward to the 1:1 line indicates the ANC difference between the median spring value and the lowest sample value for each site and year. (From Sullivan et al. 2003. *Assessment of Air Quality and Related Values in Shenandoah National Park*. NPS/NERCHAL/NRTR-03/090. U.S. Department of the Interior, National Park Service, Northeast Region. http://www.nps.gov/nero/science/FINAL/shen_air_quality/shen_airquality.html.)

fall index ANC. These kinds of relationships can be useful in evaluating the likelihood of experiencing extreme values that exceed various response thresholds for expected biological effects.

## 5.4   CONDUCT EXPLORATORY ANALYSES

### 5.4.1   Analysis of Water Quality Status

Various graphical and statistical methods are available for describing ambient water quality and assessing differences in water quality across a study area or region. Current status of water quality should be compared among sites, with previously obtained data for individual sites, with criteria values or standards used in water quality assessments, and with values that represent ecological thresholds.

There is no standard procedure for the statistical analysis of water quality data for the purpose of evaluating sensitivity to, or effects from, atmospheric deposition. Rather, there exists a range of options for depicting results or analyzing differences over space or time. Selection of methods will depend on a host of issues, including project objectives; the quantity and quality of the data; number of sampling locations; length of the period of record; extent to which samples were collected across years, seasons, and hydrological episodes; and specifics of the resulting dataset. Important dataset issues include the presence and abundance of extreme outliers, censored data, and negative values; normal versus nonnormal distribution of the data; seasonality and episodicity of the data; and extent to which data values are missing or are less than reporting limits. It is generally advisable to consult with a statistician, or an individual who is knowledgeable about statistics, prior to conducting trends analyses and other complex statistical analyses. Nevertheless, there are some commonly used and accepted data analysis approaches and statistical tests that are often applicable to the types of data analyses needed in routine water quality studies. These are described in the sections that follow.

### 5.4.2   Graphing and Qualitative Analysis

Graphics used to visualize water quality data include scatterplots of values for single or multiple sites by date. Water quality should also be examined relative to continuous variables such as elevation, watershed area, or discharge. The range and distribution of data for different periods of time or for different lakes or streams can be depicted with histograms (Figure 5.4) or box-and-whisker plots (Figure 5.6). The box plot graphically represents the central tendency and variability in a dataset. The range indicated by the box (top to bottom) represents the middle half of the data and is bisected by a line that represents the

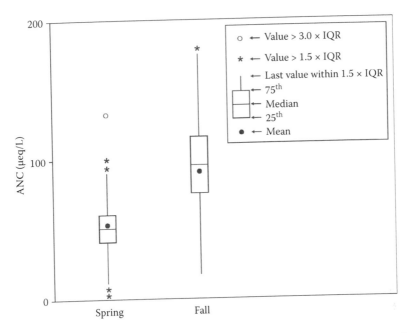

**Figure 5.6**  Example box plots to compare hypothetical lake ANC values measured in samples collected during the spring versus the fall season. IQR indicates the interquartile range, or the difference between the 25th and 75th percentile values.

median value of the data. Because the bottom of the box represents the lower quartile (25th percentile) of the data and the top of the box represents the upper quartile (75th percentile) of the data, the vertical length of the box represents the interquartile range (IQR) of the data. The end of each whisker represents the last value from the dataset that is no more than 1.5 times the IQR. The outliers (values beyond 1.5 times the IQR) are all shown on the plot. Data points marked with a star are greater than 1.5 times but less than 3.0 times the IQR and are considered possible outliers; those that are marked with an open circle are greater than 3.0 times the IQR and are considered probable outliers.

Graphics should be used to examine temporal variation in data for individual sites, including, for example, patterns associated with season or discharge, as well as gradual or more sudden changes in values. Spatial variation among multiple sites can also be represented graphically, for example including variation related to differences in watershed properties, land use, or exposure to pollutants. Even when a more quantitative statistical analysis of water quality data is desired, qualitative visual data examination is recommended as a preliminary step.

The steps that one should take in analyzing the dataset will depend to a large degree on the specifics of the dataset itself and the purpose of the analysis.

Common management issues that involve analysis of surface water field-sampling data are outlined in Table 1.3 of the study design provided in Section 1.1.2. That section of this book identified six major kinds of approaches (each tied to a purpose) as follows:

1. Determine whether one lake or stream, or a group of lakes or streams, is N limited for algal growth.
2. Quantify episodic excursions from base flow conditions in surface water chemistry during hydrologic events.
3. Determine the distribution of lake or stream water chemistry across a particular study area.
4. Quantify long-term changes in lake or stream ANC (or other variable) over time in a particular lake or stream.
5. Determine to what extent air pollution is currently affecting the water resources in a particular forest or wilderness.
6. Evaluate whether the current condition of acid- or nutrient-sensitive waters warrants mitigation.

The analyses that could, or should, be conducted will depend in part on which approach is required to answer particular management questions.

Every dataset will offer its own challenges and, if sufficiently examined, will reveal its own, often unique, patterns. Regional differences are important. Furthermore, water quality data analysis is exploratory in nature. To properly analyze a water quality dataset, the analyst must experiment with different approaches and eventually find some that work with that dataset and those specific analysis objectives.

Despite these difficulties and the site specificity of water quality data analysis, it is possible to offer recommendations and examples of steps to be considered. A successful analysis will develop through trial and error. The example analyses illustrated in this part of the DAP show some of the approaches that we have found to be useful. You may find some of these examples to be successful in some cases. Nevertheless, an analyst should always explore multiple options to determine what works best for a particular dataset. If the data are high quality, it is likely that they will tell a story. Some creativity may be required to reveal that story.

## 5.4.3   Recommended Data Analyses

We recommend various types of data analysis here. These recommendations are specific to the purpose of the data analysis as outlined in Chapter 1. You may find alternative approaches to be as successful as or more successful than those provided here. There is no one clear choice of how to approach exploratory data analysis.

### 5.4.3.1 Creating Data Subsets

Exploratory data analyses should be conducted using all of the available data. In addition, however, it is often helpful to create various subsets of the data and analyze them individually. This is because inherent variability can obscure the patterns that might exist in the particular subset of the data that represents the more sensitive or affected bodies of water, times of year, hydrological conditions, geological settings, and so on. Therefore, you should explore various ways to create your data subsets prior to conducting exploratory analyses to determine if some patterns are only evident, or are strongest, for one or more subsets, as compared with the dataset as a whole.

Data subsets can be created for exploratory analysis using water chemical criteria (Table 5.5). Alternatively, or in combination, data subsets can be created using features of the landscape, hydrology, or morphology (Table 5.6).

TABLE 5.5   EXAMPLE VARIABLES FOR CREATING WATER QUALITY DATA SUBSETS, ACCORDING TO MEASURED WATER CHEMISTRY, PRIOR TO ANALYSIS

| Variable | Possible Cutoff Values for Designating Lake or Stream Classes |
|---|---|
| ANC | 0, 20, 50, 100 µeq/L |
| $NO_3$ | 5, 10, 15 µeq/L |
| DOC | 200, 400, 500 µM |
| pH | 5.0, 5.5, 6.0, 6.5 |
| $Al_i$ | 2, 7 µM |
| Ca + Mg[a] | Highly region specific |
| $SO_4$ | Highly region specific |

[a] Can be analyzed combined or individually; in some cases (where they are quantitatively important), Na or K might also be included.

TABLE 5.6   EXAMPLE VARIABLES FOR CREATING WATER QUALITY DATA SUBSETS PRIOR TO ANALYSIS ACCORDING TO FEATURES OTHER THAN MEASURED WATER CHEMISTRY

| Variable | Possible Lake or Stream Classes |
|---|---|
| Geologic class | For example, siliciclastic, granitic, argillaceous, and so on |
| Elevation | Can use above or below a specific cutoff or as discrete elevational bands |
| Lake type | Drainage, seepage, type of seepage lake (perched or flow-through) |
| Stream Strahler order | Can combine into classes (i.e., 1st plus 2nd, 3rd plus 4th, etc.) or analyze as individual orders |

This is an opportunity for the analyst to be creative. Try different approaches and see what works. The objective is to improve your understanding of the data and the story that they have to tell.

### 5.4.3.2 Determine Whether One Lake or Stream, or a Group of Lakes or Streams, Is N Limited for Algal Growth

Productivity of surface water can be limited by a multitude of factors. For example, small streams are commonly limited by light; if the stream is highly shaded by riparian vegetation, then primary productivity may be low even if nutrient concentrations are high enough to support extensive algal growth. Streams can also be limited by substrate type. For example, if suitable substrate is not available for attachment, then algal productivity may be low relative to nutrient availability. Streams and lakes can also be limited by a nutrient. This is most commonly P or N. If N is limiting, then atmospheric contributions of N can enhance productivity, contribute to eutrophication, and perhaps alter species composition and abundance.

The relative importance of N and P, as potentially limiting or colimiting nutrients, can be evaluated by conducting a rough screening analysis based on the molar ratio of N:P concentrations in surface water. This ratio ideally should include all forms of N and P, both particulate and dissolved, both organic and inorganic. Thus, total N is the sum of the concentrations of $NO_3^--N$, $NH_4^+-N$, and organic N. [Note that $NO_3^--N$ refers to the portion of the $NO_3^-$ that is comprised of N and not by O; $NH_4^--N$ reflects the portion of the $NH_4^+$ that is comprised of N.] Phosphorus is measured as total P. Units are in micromoles per liter for both elements. If measurements of total N are not available, one can use an estimate of total inorganic N (TIN), calculated as the sum of the molar concentrations of $NO_3^-$ and $NH_4^+$. Note that the concentrations of $NO_3^--N$ in micromoles per liter are the same as the concentrations in microequivalents per liter; no conversion is needed. Total N in micromoles per liter is equal to the concentration of total N in milligrams per liter times 71.38. Total P in micromoles per liter is equal to the concentration in milligrams per liter times 32.29.

Based on available experimental data, a large majority of lakes that have the ratio of total N to total P below about 44 have been found to indicate N-limited phytoplankton growth (Elser et al. 2009). This is an area of active research, and interpretations may be subject to change in the near future. Lakes and streams that are nutrient (N or P) limited, rather than by light or some other factor, may change in their limitation status over time, perhaps with season. Therefore, temporal patterns in the N:P molar ratio could be examined over time. In addition, we recommend evaluation of spatial patterns in N:P to determine, for example, if water bodies in a study area tend to be N limited primarily at certain elevations, on certain geological types, or in certain vegetation communities.

Thus, N:P ratios could be mapped relative to landscape condition to reveal such patterns, if they occur.

Determination that a water body, or a group of water bodies, exhibits potential N limitation based on the N:P ratio is not sufficient evidence to indicate that the system is, in fact, N limited. Furthermore, such an analysis does not necessarily identify any variability that might occur in that nutrient status throughout the growing season. These recommended N:P ratio analyses are screening-level analyses that may suggest the possibility of N limitation.

The next step in the assessment process is to conduct laboratory studies to determine N versus P limitation. These could involve collection of multiple liters of lake water, which is then shipped on ice to the laboratory. The water is dispersed into flasks, typically at least three flasks per treatment. Treatments may involve multiple light levels and varying (low and high) nutrient additions of N only, P only, N plus P, and control (no nutrient addition). Incubation, with continuous or periodic mixing, is conducted under approximate ambient lake temperature conditions. Algal growth can be tracked daily by measuring the chlorophyll *a* concentration in an aliquot of water from each flask.

Based on the results of the laboratory incubation studies, and the degree of rigor required for the project in the determination of nutrient limitation, it may be desirable to progress to *in situ* incubation studies. Such experiments should involve *in situ* incubation of water over a period of time during the growing season in multiple containers suspended in the lake or stream. The containers should include a control (no nutrient addition), multiple (at least two levels: high and low) N addition containers, multiple P addition containers, and multiple N-plus-P addition containers. For example, one may double and triple the ambient nutrient concentrations in the two containers for each type of nutrient input. Changes in the concentration of chlorophyll *a* over time in the treatment containers, relative to the control, indicate productivity responses to nutrient addition. Such experiments can verify whether, and when, a lake or stream (or a group of lakes or streams) may be susceptible to eutrophication effects associated with atmospheric N deposition.

### 5.4.3.3   Quantify Episodic Excursions from Base Flow Conditions in Surface Water Chemistry during Hydrologic Events

Changes in the concentrations during episodes of major ions, pH, and ANC should be evaluated for a given lake or stream by plotting individual measured values during multiple storms. An example for one lake or stream during one storm or snowmelt episode is shown in Figure 5.7.

The extent to which ANC and pH decrease and the extent to which $Al_i$ increases in response to hydrological episodes provides an indication of chemical extremes to which aquatic biota are exposed during hydrological episodes.

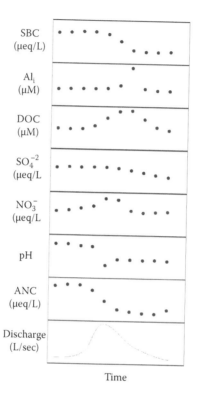

**Figure 5.7**   Changes in the concentration of major water chemistry constituents during a hypothetical hydrological episode in one stream. SBC is the sum of the four base cation concentrations. Data for each variable of interest are plotted along the same time axis and compared at the same scale relative to the pattern of discharge.

Patterns of episodic responses of sum of base cations (SBC), DOC, $SO_4^{2-}$, and $NO_3^-$ concentration can reveal important information regarding the causes of episodic excursions of ANC, pH, and $Al_i$. Both ANC and pH can decrease, and $Al_i$ can increase, in response to base cation dilution (decreased SBC), $NO_3^-$ leaching, $SO_4^{2-}$ leaching, and DOC mobilization. The relative importance of these various potential drivers varies by watershed, by region, by season, and by hydrologic event. Examination of the kinds of plots shown in Figure 5.7 can reveal these patterns in the various potential drivers at one site during one event. It may be necessary to sample and analyze multiple sites and multiple events.

Similarly, temporal patterns of changing water chemistry in a given lake or stream should be examined across the annual or seasonal cycle. In regions having marked snowpack development, such an analysis should include the entire snowmelt period, as is shown in Figure 5.8. Intensive time series data, where available, provide finer resolution of episodic changes in chemistry.

An analysis of surface water $NO_3^-$ concentration as a fraction of the total combustion-related mineral acid anion ($SO_4^{2-}$ plus $NO_3^-$) concentration can reveal the relative roles of $NO_3^-$ and $SO_4^{2-}$ in influencing surface water chemistry. In many surface waters, under certain hydrological conditions,

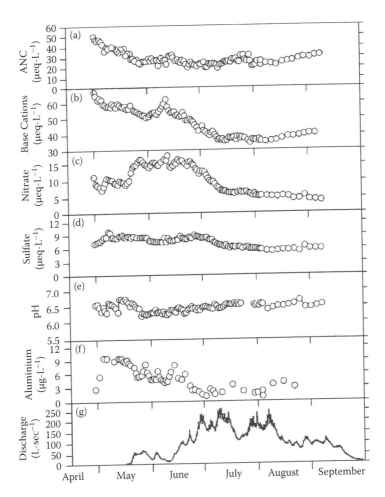

**Figure 5.8**   Time series of major ions and discharge in Treasure Lake in the Sierra Nevada during snowmelt in 1993. Seasonal and episodic changes in surface water chemistry can be examined using these simple time series plots. (Reprinted, with permission, from Stoddard, J.L. 1995. *Water Air Soil Pollut.* 85:353–358. Copyright 1995 Springer.)

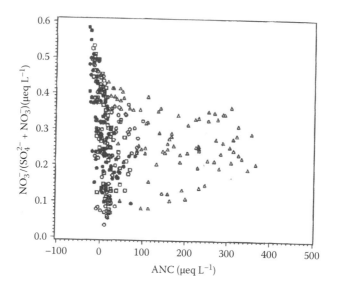

**Figure 5.9**   Ratio of $NO_3^-$:$(SO_4^{2-} + NO_3^-)$ concentration versus ANC in stream water samples collected during hydrological episodes in four streams included in the Adirondack region of EPA's Episodic Response Program (ERP). The different symbols on the graph represent different streams. (Reprinted, with permission, from Sullivan et al. 1997. *Water Air Soil Pollut.* 95(1–4):313–336. Copyright 1997 Springer.)

$SO_4^{2-}$ concentration largely determines the total mineral acid anion concentration. When this occurs, $NO_3^-$ has relatively little influence on the ANC or pH of the water. There can, however, be times when $NO_3^-$ is also important to the acid-base status of the water. For example, data collected from four Adirondack Mountain streams during hydrological episodes (Figure 5.9) illustrated that, for the study streams, (1) $NO_3^-$ generally provides less than half of the contribution of mineral acid anions from the atmosphere ($SO_4^{2-}$ provides the majority), but (2) the relative importance of $NO_3^-$, compared to $SO_4^{2-}$, increases at lower ANC values (which occur during high-flow periods).

### 5.4.3.4   Determine the Distribution of Lake or Stream Water Chemistry across a Particular Study Area

Patterns in water chemistry should be mapped to illustrate spatial patterns in the data. Figure 5.10 shows one way to do that, in this case for lake water $NO_3^-$ concentration in Adirondack lakes. Each bar represents one lake; the base of the bar reflects the lake location. The height of each bar is proportional to the $NO_3^-$ concentration. In this example, concentrations of $NO_3^-$ are highest in the southwestern portion of the Adirondack park and in the central high peaks area, the general locations where N deposition and precipitation

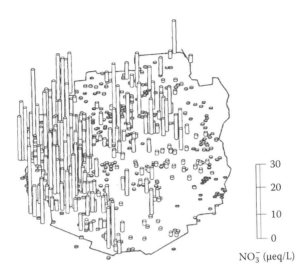

$NO_3^-$ (μeq/L)

**Figure 5.10**  Map of summer $NO_3^-$ concentrations in drainage lakes sampled by the Adirondack Lakes Survey Corporation in the Adirondack region of New York. Maps such as this can reveal spatial patterns in the concentration of any surface water variable across a study area. (Reprinted, with permission, from Sullivan et al. 1997. *Water Air Soil Pollut.* 95(1–4):313–336. Copyright 1997 Springer.)

amounts are highest. Thus, spatial patterns in surface water chemistry could be compared with various factors that are known or suspected to be associated with water chemistry. These might include geology, soil types, elevation, atmospheric deposition, precipitation amounts, vegetation types, and so on.

Spatial patterns can also be analyzed across a gradient of deposition or across a gradient of expected resource sensitivity using space-for-time substitution analysis. In this approach, it is assumed that the lakes or streams across the study area were initially relatively homogeneous in their chemistry, and furthermore that differences observed across space at the present time correspond with changes that occurred in the past. Such an analysis could be conducted across a gradient in elevation, deposition, slope steepness, and so on rather than, or in addition to, across a gradient in deposition. The slopes of the data depicting changes across spatial gradients provide estimates of the quantitative importance of the various changes. Such quantitative estimates can be combined with model estimates of changes at selected locations in a weight-of-evidence assessment.

Portions of a study area can sometimes be identified and mapped within which most of the acid-sensitive or nutrient-sensitive waters are expected to occur. In the example from the Southern Appalachian Mountains Initiative (SAMI) study shown in Figure 5.11, the area delimited by an acidification sensitivity classification scheme is shown. The darkly shaded area includes the siliceous geo-logic sensitivity class surrounded by a 750-m buffer. In addition, all areas less than 400-m elevation have been deleted, and areas greater than 1000-m elevation have been added. The area thus circumscribed includes 95% of the known acidic streams and 88% of the known streams having ANC of 20 µeq/L or less (of more

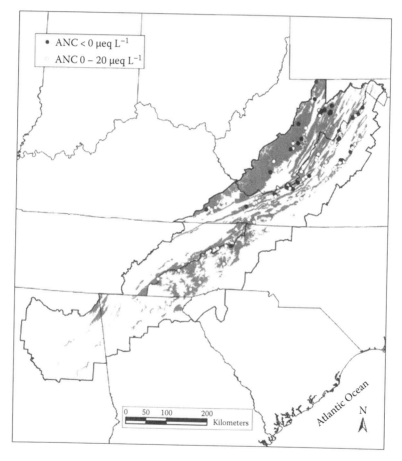

**Figure 5.11** Map showing the results of a classification system devised to reveal the locations where low-ANC streams were expected to occur in the southern Appalachian Mountains (in this example, based on geology and elevation) compared with the locations of all streams known to have low ANC, which are represented as dots on the map. Streams having ANC less than 0 µeq/L are coded in black; streams having ANC between 0 and 20 µeq/L are coded in white. (Reprinted, with permission, from Sullivan et al. 2007. *Water Air Soil Pollut.* 182:57–71. Copyright 2006 Springer.)

than 900 streams surveyed) within the region. Furthermore, all known streams having ANC equal to 20 μeq/L or less are in close proximity to the final mapped area. In some cases, you can use an approach such as this to circumscribe portions of a study area thought to contain most of the sensitive or impacted water bodies.

The distribution of data values across a given study region can reveal important information about the source of a constituent. For example, it can be helpful to plot the frequency distribution of surface water $SO_4^{2-}$ concentrations within a relatively small study area. Atmospheric S contributions to watersheds are expected to yield a reasonably well-defined bell-shaped or half-bell-shaped curve in surface water $SO_4^{2-}$ concentrations. Differences from one study watershed to another in such features as elevation, aspect, vegetation type, and topography contribute to variability, but the overall patterns should be relatively homogeneous if the study area is relatively homogeneous. The observed outlier lakes or streams having much higher concentrations of $SO_4^{2-}$ than the population of lakes or streams at large can be presumed to receive contributions of geological S unless there is a good reason why atmospheric deposition should be markedly higher at those outlier locations. Lakes or streams that contain appreciable geological S are not good candidates for monitoring or study to quantify effects from atmospheric sources of S.

Spatial patterns in water chemistry or landscape characteristics can also be used to aid in extrapolating results from a relatively few intensively studied sites to the larger region. In many cases, model simulations of future chemistry or critical deposition load may be available for only a small subset of the lakes or streams in a given region. Such results can sometimes be extrapolated to the wider population of waters using relationships with water chemistry (such as, for example, ANC in the example shown in Figure 5.12) or landscape features that correlate with sensitivity.

### 5.4.3.5 Quantify Long-Term Changes in Lake or Stream Chemistry over Time

Detection of trends in water quality over time can be complicated by analytical error and measurement uncertainty that contribute to scatter in time series data. Plus, inter- and intraannual variability contribute additional scatter. In particular, seasonal and episodic variability (which are largely driven by climate and hydrology) often contribute to short-term changes in water chemistry caused by acidification or nutrient N addition (Figure 5.13). Therefore, it can be expected that many years of monitoring data may be needed to reveal a probable trend (we recommend at least 8 years). In general, it is helpful to collect data during multiple times of each year to reveal the variability that exists and to discern the trend that exists within the

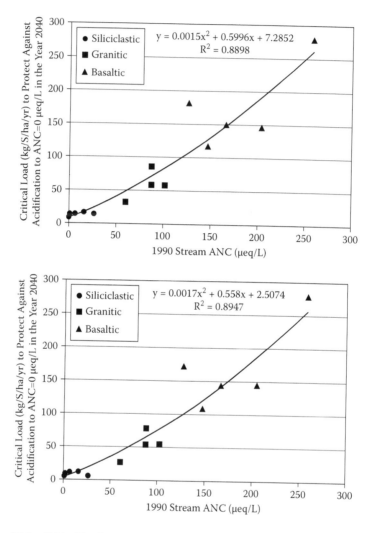

**Figure 5.12**  Critical load simulated by the Model of Acidification of Groundwater in Catchments (MAGIC) to protect streams in Shenandoah National Park against acidification to ANC below 0 (top panel) and 20 µeq/L (bottom panel) by the year 2040 is plotted as a function of 1990 ANC. Stream sites are coded to show differences in geology. This approach yields a predictive equation with which to estimate the model projected value (in this example of critical load) for a specific stream based on the measured value of ANC in that stream. (Reprinted, with permission, from Sullivan et al. 2008. *Environ. Monitor. Assess.* 137:85–99. Copyright 2007 Springer.)

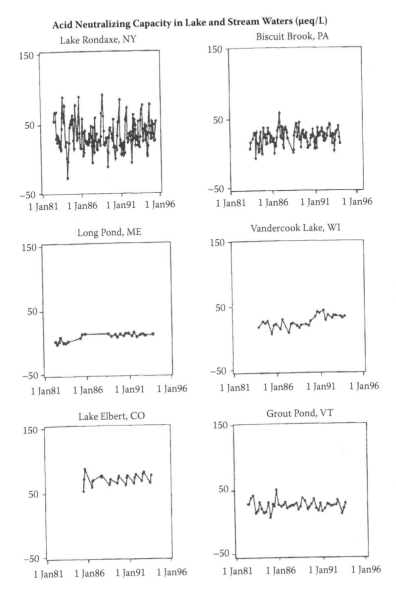

**Figure 5.13**   Some example trend analyses of ANC in lake and stream waters, based on data from EPA's Long-Term Monitoring (LTM) program. These examples illustrate that interannual and intra-annual variability can sometimes be larger than the change over time in the variable of interest. (Reprinted, with permission, from Sullivan, T.J. 2000. *Aquatic Effects of Acidic Deposition*. Lewis/CRC Press, Boca Raton, FL. Copyright 2000 CCC Republication.)

noise of that variability. In other cases, it can be helpful to standardize time series data to minimize the influence of hydrological differences in constituent concentrations. The purpose is to minimize or eliminate the variability associated with seasonality in the data. This can be accomplished in multiple ways, including:

- focusing on summer or fall index chemistry, with collection of samples under conditions having minimal influence of snowmelt or rainstorm events;
- representing the available data as discharge-weighted average values if discharge is available for the site in question or can be estimated, or indexed, from a nearby site that is gauged; or
- focusing only on the minimum (i.e., ANC) or maximum (i.e., Al$_i$) concentration measured during each year; in this case, the plot would be of, for example, the lowest (of multiple measurements each year) measured ANC value each year over a period of 8 or more years.

### 5.4.3.6  Determine to What Extent Air Pollution Is Currently Affecting the Water Resources in a Study Area

Is air pollution affecting water quality and if so, by how much? There are many ways to analyze data to shed light on these questions. In general, no single approach should be considered definitive. When multiple approaches converge to provide similar conclusions, there is greater confidence in the validity of that conclusion. Some of the figures presented previously (e.g., Figures 5.7 through 5.11) can provide useful information. Another approach entails plotting the relationships between the ratio of $SO_4^{2-}$ (or $SO_4^{2-} + NO_3^-$) to the sum of the base cation (SBC) concentrations versus ANC. A clear pattern across sites of decreasing ANC as the ratio of mineral acid anion (presumed to have been derived from acidic deposition) to base cation (reflective of ecosystem acid buffering) increases suggests that ANC has decreased in response to acidic deposition. The concentration of DOC can alter the relationship by decreasing ANC below what would otherwise be expected at a given $SO_4^{2-}$/SBC ratio. Streams or lakes in some regions tend to have uniformly low DOC, so creating subsets on DOC concentration is not necessary. Because $NO_3^-$ concentration is relatively high at stream sites in some regions, the ratio can include $NO_3^-$ and be presented for that region as $[SO_4^{2-} + NO_3^-]$:SBC.

Assessments of acidic deposition effects and recovery generally rely on ANC and pH as the primary chemical indicators. However, both of these measurements can be influenced by naturally produced organic acidity associated with DOC, which can be abundant in streams and lakes draining wetlands and to a lesser extent coniferous forests. In waters with significant concentrations of DOC, acidity from acidic deposition can be distinguished from natural organic acidity using the base cation surplus (BCS). The BCS is an index that is based on the mobilization of toxic inorganic aluminum within the soil. In the absence of acidic deposition or geological S, inorganic Al remains in the soil in a non-harmful form. However, acidic deposition dissolves soil Al in a form that moves from soils into surface waters and harms both terrestrial (Minocha et al. 1997, Long et al. 2009) and aquatic life (Baldigo et al. 2007, Lawrence et al. 2008a, Baldigo et al. 2009). A BCS value less than 0 µeq/L in surface water generally indicates that the soil has become sufficiently acidified by acidic deposition to enable toxic forms of aluminum to be mobilized (Lawrence et al. 2007, 2008b). A negative BCS value could also occur from acid mine drainage or where drainage waters pass through geologic deposits rich in sulfide-bearing minerals. A BCS value between 0 and 50 µeq/L indicates a watershed with low calcium availability, which (1) is at risk of future acidification from continued acidic deposition and (2) can limit the productivity of aquatic (Jeziorski et al. 2008) and terrestrial (Long et al. 2009) ecosystems.

The BCS can be calculated using variables typically measured in low-ionic-strength waters at risk from acidic deposition:

$$BCS = (Ca^{2+} + Mg^{2+} + Na^+ + K^+) - (SO_4^{2-} + NO_3^- + Cl^- + RCOO^-_s) \quad (5.12)$$

$$RCOO^-_s = 0.071(DOC) - 2.1 \quad (5.13)$$

where all concentrations used in Equation 5.12 are in milliequivalents per liter, and the concentration of DOC in Equation 5.13 is expressed in micromoles per liter.

Perhaps the most straightforward way to determine whether, and to what extent, a given lake or stream has acidified is to construct a model hindcast of past water chemistry. The two models most commonly used for such purposes are the Model of Acidification of Groundwater in Catchments (MAGIC) (Cosby et al. 1985) and Photosynthesis and Evapotranspiration–Biogeochemistry (PnET-BGC) model (Gbondo-Tugbawa et al. 2001). Each of these models has been widely used across the United States to model watershed acid-base chemistry, including hindcasts of past chemistry and forecasts of future chemistry under differing future deposition rates, and to estimate critical loads of deposition to provide resource protection or to allow damaged resources to recover.

In addition, paleolimnological reconstructions of past lake water chemistry can be constructed from the fossil remains of algal diatoms or chrysophytes (cf. Charles et al. 1990, Charles and Smol 1990) in dated lake sediment cores. If both process model hindcast simulations and paleolimnological reconstructions suggest past acidification (especially if the estimates of change are quantitatively similar), then there is increased confidence in the validity of that conclusion.

It can be helpful to discern what types of lakes or streams within a study area have the lowest ANC, highest $NO_3^-$ concentration, highest $Al_i$ concentration, and so on. Such types of waters become potentially important sites for further study or enhanced protection. For example, it may be that the lakes or streams having particularly low ANC generally, or entirely, are small. In the Adirondack Mountain region, lake ANC is related to lake area (Figure 5.14). In this example, small lakes are more likely to be both low and high in ANC; intermediate-size lakes tend to have more intermediate chemistry. Few lakes larger than 20 ha are acidic in this region. Similarly, it can be helpful to examine relationships between lake or stream chemistry and other morphometric features of the landscape, such as watershed area, stream order, lake depth, watershed slope, and so on.

Across a given region, the leaching loss of $NO_3^-$ from soil to drainage water, which can be expressed as $NO_3^-$ (sometimes also including $NH_4^+$) outputs in units of mass of N per unit watershed area per year, is related to N inputs in deposition (expressed in the same units). For example, a threshold relationship

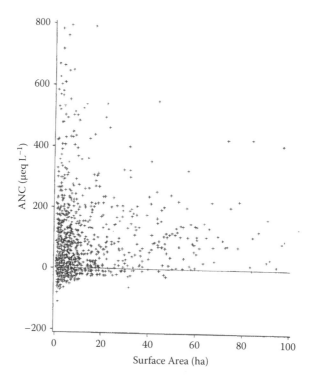

**Figure 5.14** Relationship between lake size and lake ANC in the Adirondack Mountains. (Reprinted, with permission, from Sullivan et al. 1990. *Water Resour. Bull.* 26:167–176. Copyright 2007 John Wiley and Sons.)

was shown for research sites across northern Europe (Figure 5.15). In this example, N output was low at N deposition levels below about 9 kg N/ha per year. Here, N leaching became pronounced for some, but not all, sites at N deposition above 9 kg/ha per year. In addition, N leaching became consistently high at N deposition above 25 kg/ha per year. Within a region, it might be possible to use this analysis approach to identify at what level of N deposition leaching of N to stream or lake water becomes pronounced.

Nitrogen leaching is not always governed entirely, or even mainly, by N deposition. Other factors, especially climatic factors, can also be important. For example, Moldan and Wright (1998) showed a strong relationship between N leaching and air temperature at a research site in Sweden. This analysis suggests that N dynamics at this research site might be strongly controlled by climatic condition, in this case air temperature. Precipitation or snowpack condition could similarly be important.

Nitrogen saturation of aquatic ecosystems has been described in stages, from stage 0, which reflects relatively pristine, unimpacted conditions, to stage 3, which reflects advanced N saturation (Stoddard 1994). Seasonal surface water nitrate concentration peaks at stage 0 are generally rather low (less than about 25 μeq/L) and of relatively short duration (Figure 5.16). At stage 1, the peaks in surface water $NO_3^-$ concentration are higher and the period of

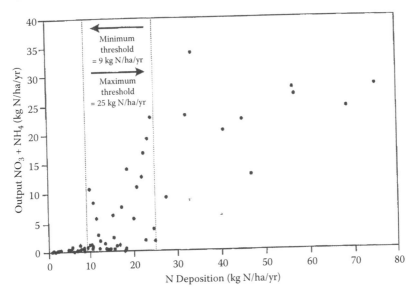

**Figure 5.15**   Nitrogen outputs in soil water or stream water versus N deposition inputs throughout Europe. (Reprinted, with permission, from Dise, N.B., and R.F. Wright. 1995. *For. Ecol. Manage.* 71:153–161. Copyright 1995 Elsevier.)

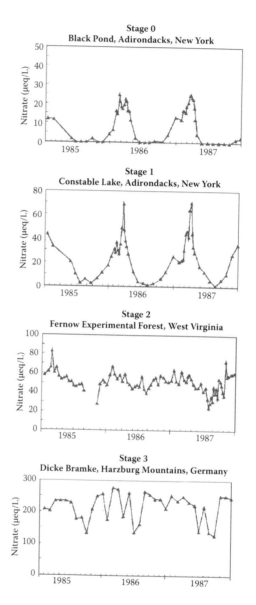

**Figure 5.16**  Example patterns of $NO_3^-$ concentration in surface water at four sites at various stages of watershed N saturation. (Reprinted, with permission, from Stoddard, J.L. 1994. Long-term changes in watershed retention of nitrogen: its causes and aquatic consequences. In L.A. Baker (Ed.), *Environmental Chemistry of Lakes and Reservoirs*. American Chemical Society, Washington, DC, pp. 223–284. Copyright 1994 American Chemical Society.)

elevated $NO_3^-$ concentration is more extensive. Under stage 2 N saturation, $NO_3^-$ concentrations remain elevated throughout the annual cycle, but some seasonality is still evident. At stage 3, the N output may actually be greater than the N input (when expressed as mass per unit area per year).

The temporal pattern of $NO_3^-$ concentration in a given lake or stream can indicate the probable stage of N saturation of the watershed. An analysis like this for a given stream or lake can illustrate the stage of N saturation of that water body, and its drainage area.

### 5.4.3.7  Evaluate Whether the Current Condition of Acid- or Nutrient-Sensitive Waters Warrants Mitigation

There is no standard analysis approach that will determine whether mitigation is warranted. This is a management judgment that should be based on a variety of analyses, as outlined previously. Potential mitigation strategies can include imposing tighter controls on atmospheric emissions of S or N and adding base cations to waters or watersheds by liming. Virtually all of the approaches suggested can contribute to such decision making.

## 5.5  CONDUCT, IF NEEDED, STATISTICAL ANALYSES

### 5.5.1  Statistical Tests for Difference

It should not be automatically assumed that formal statistical tests are needed in analyzing a dataset. Much can be gained by conducting routine exploratory data analyses, such as those outlined in the previous section, without adding the complexity of conducting formal statistical tests. In many cases, the assistance of a statistician or other person who is knowledgeable about statistics will be needed for conducing such tests.

Statistical tests are often used to determine the existence of significant differences between groups of sites or samples. The most common tests are parametric, and include $t$ tests and ANOVA that use means and variances to determine significant differences among group means. These parametric tests, however, make assumptions about data normality and independence that need to be examined before using them. Data are normally distributed if the various concentrations measured at different times for the same site or at different sites are bell shaped (Figure 5.17). If the data are not normally distributed, they must be transformed before parametric analysis, or they must be analyzed using a nonparametric test. For water chemistry, variables are often log transformed to achieve a normal distribution. There are a number of statistics that can test for normality (e.g., Kolmogorov-Smirnov test), but we recommend, instead of applying such tests, plotting histograms or some other type of frequency plot

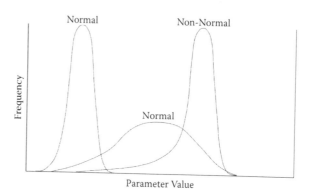

**Figure 5.17**   Schematic representation of data normality.

and visually inspecting the graph to see if the distribution deviates grossly from a bell-shaped normal distribution.

Nonparametric or distribution-free statistics are used to test for group differences in skewed (not normally distributed) data or when the analyst is not comfortable with assumptions about normality. Water chemistry data are commonly *not* normally distributed. The nonparametric tests are based on sample ranks (rank order number from low to high) and not actual data values. Because the test is based on rank, no assumption is made about the underlying data distribution. The best-known nonparametric test for group difference is the Mann-Whitney U test or, as it is also known, the Wilcoxon rank sum test.

Nonparametric tests, including the rank sum test, tell you nothing about the magnitude of the difference between groups, just whether the group differences are significant. Analysts should be cautioned, however, that with large enough sample sizes, groups can be statistically different but such differences may have little ecological significance. The magnitude of any revealed differences need to be examined. Just running a statistical test to determine if differences are statistically significant is not sufficient. It only tells part of the story.

There are multiple forms of the rank sum test, and the choice of which form to use is complex (Helsel and Hirsch 1992). We recommend that this test not be applied by persons lacking formal training in statistics. The rank sum test determines whether one group of measurements tends to produce higher values than another group of measurements. In other words, the test determines if both groups of data are from the same population. The groups might represent different lake types, different periods of time, different seasons, and so on.

For comparing more than two independent groups of data points, the Kruskal-Wallis test is often used. It can be computed by an exact method used for small sample sizes (typically five or fewer samples per group), a large-sample (chi-square) approximation, or ranking the data and performing a parametric test on the resulting ranks (Helsel and Hirsch 1992). The last two methods only produce valid $p$ values when sample sizes are large. The null hypothesis for all variations of this test specifies that all of the groups have identical data distributions (or have the same median value); the alternate hypothesis specifies that at least one group differs from the others with respect to its data distribution or its median value.

Just as for the rank sum test, all observations are combined together and ranked from lowest (1) to highest ($n$). The average group rank $R\phi_j$ is compared to the overall average rank to calculate the test statistic. As for the rank sum test, we recommend that this test should be conducted by a person who has had formal training in statistics.

## 5.5.2   Trend Detection

Trends in water chemistry over time can be evaluated using simple linear regression (SLR) or using a more sophisticated statistical approach, such as the nonparametric seasonal Kendall tau (SKT) test (Hirsch and Slack 1984) for determining monotonic trends in seasonally varying water quality. The SLR approach is simpler to apply and will sometimes yield nearly identical estimates of slope as the SKT test (Sullivan et al. 2003). An example SLR analysis for $SO_4^{2-}$ concentration in Deep Run, a small acid-sensitive stream in Shenandoah National Park, is shown in Figure 5.18. The slope of the regression

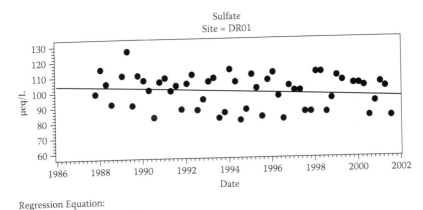

Regression Equation:
$SO_4 = 1244.275 - 0.573908*Date$

**Figure 5.18**   Plot, with regression line, of $SO_4^{2-}$ concentration in Deep Run, Shenandoah National Park, over the period of monitoring record through 2002. (From Sullivan, et al. 2003. *Assessment of Air Quality and Related Values in Shenandoah National Park.* NPS/NERCHAL/NRTR-03/090. U.S. Department of the Interior, National Park Service, Northeast Region. http://www.nps.gov/nero/science/FINAL/shen_air_quality/shen_airquality.html.)

(−0.57 μeq/year) indicates the rate of change (in this case, a decrease) in the variable ($SO_4^{2-}$) over time. Whether the relationship between $SO_4^{2-}$ concentration and time is statistically significant can be determined using the $p$ statistic. We recommend a $p$ value of less than or equal to .05 as the benchmark for determining statistical significance. The $r^2$ statistic can be used to determine the percentage of the variation in $SO_4^{2-}$ concentration that is explained by the variable time.

If the regression relationship is statistically significant at $p$ of .05 or less, the next step is to conduct a test to determine if the slope of that relationship is significantly different from zero. If it is determined that the slope is statistically either greater or less than zero, then it can be concluded that the parameter in question is truly increasing or decreasing over time.

The test statistic for determining if the slope is different from zero is expressed as

$$t = \frac{r\sqrt{n-2}}{\sqrt{1-r^2}} \tag{5.14}$$

where $r$ is the correlation coefficient of the regression, and $n$ is the number of data points. The null hypothesis that the slope of the regression equals zero is rejected if the absolute value of t is higher than $t_{crit}$, where $t_{crit}$ is the point on the Student $t$ distribution with ($n$ minus 2) degrees of freedom and with a probability of exceedance of $\alpha/2$. The Student $t$ distribution for that CL is given in Table 5.7.

In general, we recommend conducting SLR analyses as the routine approach to determine changes in water chemistry over time. If it is deemed necessary to obtain a more rigorous trends estimate, then a more complex statistical analysis can be conducted. In some cases, there may be a visually obvious change in the slope of the data points in the midst of the time series. For example, some water bodies in the United States experienced acidification during the 1980s and 1990s, but recovery (decreasing ANC) is evident after about the late 1990s. In such a situation, it can be helpful to visually split the data into two time periods (acidification and recovery) and perform an SLR separately on each time period to determine if there has been a change over time.

Simple linear regression analysis is a good first approach for analyzing temporal data and can be performed using common spreadsheet and statistical software. However, regression analysis is sensitive to data normality and other assumptions and is often not the most robust method to quantify the statistical significance of temporal trends. Loftis et al. (1989) evaluated a number of different trend detection methods under a number of different conditions and found that there is no one method that outperforms the others under all conditions. However, they found that the most

**TABLE 5.7    VALUES OF $t$**

| $df$ | Probability ($\alpha$ = 0.05) of a Numerically Larger Value of $t$ |
|---|---|
| 1 | 12.706 |
| 2 | 4.303 |
| 3 | 3.182 |
| 4 | 2.776 |
| 5 | 2.571 |
| 6 | 2.447 |
| 7 | 2.365 |
| 8 | 2.306 |
| 9 | 2.262 |
| 10 | 2.228 |
| 11 | 2.201 |
| 12 | 2.179 |
| 13 | 2.160 |
| 14 | 2.145 |
| 15 | 2.131 |
| 16 | 2.120 |
| 17 | 2.110 |
| 18 | 2.101 |
| 19 | 2.093 |
| 20 | 2.086 |
| 21 | 2.080 |
| 22 | 2.074 |
| 23 | 20.69 |
| 24 | 2.064 |
| 25 | 2.060 |
| 26 | 2.056 |
| 27 | 2.052 |
| 28 | 2.048 |
| 29 | 2.045 |
| 30 | 2.042 |

*Continued*

| TABLE 5.7 (*Continued*) | VALUES OF *t* |
|---|---|
| *df* | **Probability (α = 0.05) of a Numerically Larger Value of *t*** |
| 40 | 2.021 |
| 60 | 2.000 |
| 120 | 1.980 |
| ∞ | 1.960 |

*Note:* Probability ($\alpha/2 = .025$) of a larger positive value of *t*.

powerful methods under most conditions were nonparametric tests that looked at the correlation between rank order and time. For annual data, the Kendall tau test (also called the Mann-Kendall test for trend) was generally the most powerful. For seasonal data, the SKT and seasonal analysis of covariance (ANCOVA) on ranks were the most powerful tests. These nonparametric tests performed about as well as parametric tests with normal data and outperformed them when the data were nonnormal. We recommend their use for testing the statistical significance of temporal trends if a more rigorous statistical application than SLR is desired. Note that these tests are the most powerful for testing whether water quality is changing over time. They do not, however, quantify the magnitude of change. Trend detection in surface waters can be conducted using complex statistical tests that attempt to adjust for natural variation related to seasonality and variations in discharge. Different methods have been developed for assessing trends or the effect of time, with proper method selection dependent on assumptions related to the distribution and independence of the data and on whether change occurs monotonically or as a step change. It is also possible to construct complicated models that incorporate flow, temperature, or other environmental factors in addition to the time variable to quantify trends. These regression-based approaches, however, are sensitive to issues of data normality and independence. In addition, different methods are available for assessing trends for individual streams or lakes, as well as for assessing regional trends associated with classes or populations of streams and lakes. These methods were described by Helsel and Hirsch (1992), Stoddard et al. (2003), and Irwin (2008). In general, we do not recommend application of such tests for routine water chemistry assessment.

If a more complicated test is to be conducted, trends in time series data collected at quarterly, monthly, or weekly intervals for individual surface waters are most commonly assessed using the SKT test developed and described by Helsel and Hirsch (1992). The nonparametric SKT is based on the correlation between the ranks of the dependent variable (concentration) and an evenly spaced time interval. The SKT is popular because of its relative simplicity compared to other approaches and minimal data assumptions. It is appropriate for data showing seasonal cycles, and it is robust with respect to issues of normality, missing or censored data, and serial correlation. It can be applied on unadjusted chemical concentration data or on residuals from ordinary least-squares regression of concentration on estimated discharge, thus accounting for the effects of changes in discharge. An alternative approach is to remove the seasonality from the dataset (e.g., by creating subsets to only include samples collected during summer base flow) and then analyzing the data in the subsets as an annual, rather than seasonal, dataset. The SKT provides the significance and direction of any trend. A different test based on the median of the set of slopes calculated for all possible pairs of points in the time series is commonly applied to calculate the slope or rate of change associated with the overall trend (Sen 1968).

Regional trends associated with classes or populations of surface waters can be assessed using a median trend test Statistical Analysis System (SAS) Institute 1988, Altman et al. 2000, Stoddard et al. 2003). This test, a meta-analysis, is based on the median of slopes obtained for linear regression of concentrations with time for the individual surface waters in the class or population of concern. Regional trend significance is tested by estimating confidence intervals around the median values, with median slopes significantly different from zero taken to indicate regional increasing trends (positive median slope) or regional decreasing trends (negative median slope). This test allows determination of trends for a resource management or geographic unit as a whole. It can be applied to include multiple predictor variables in the regression models, thus accounting for other factors, such as discharge, in addition to time.

### 5.5.3  Statistical Power

For purposes of inferring differences or change, data analysts and resource managers will need to make decisions concerning acceptable error levels, and such decisions should be based on the allocation of risk, given the relative importance of ensuring resource protection compared to the cost of potentially unnecessary responses to perceived damage. These decisions are typically made in terms of statistical power and significance. Statistical power refers to the avoidance of false negatives or wrongly concluding that a change or impact has not occurred when, in fact, a change or impact has occurred (type II error). Statistical significance refers to the avoidance of false positives or wrongly concluding that a change or impact has occurred when, in fact, no change or impact has occurred (type I error). Statistical power and significance levels are stated as percentages. A statistical power level of 90%, for example, would mean that 90% of the time an effect of a specified size (whatever it is) will be correctly identified. A statistical significance level of 5%, for example, would mean that only 5% of the time would an identified effect be incorrectly identified (or there is a 95% probability that the identified effect is, in fact, correctly identified). Ideally, statistical power and significance objectives are established as part of the initial study or monitoring program design and not during data analysis. They will be stated in terms of the project's ability to detect a specific effect or a specified trend (a magnitude of change within a specified time period).

The ability of a monitoring program to detect temporal trends is a function of a number of different factors. For detecting a trend at a single site, for a specified type I and type II error rate (e.g., a 95% CL), trend detection is determined by the magnitude of the actual trend you wish to detect, how long you have to detect the trend, and the variability of the water quality parameter assessed. For water chemistry assessment, trend magnitude is usually expressed as a percentage change in the variable of interest per year. Evaluation of the magnitude of the trend one wishes to be able to document is usually based on program objectives related to ecologically significant changes in water quality values within a time period of policy relevance. There is no standard procedure for this, but in general terms one typically wishes to be able to document for acid-base chemistry monitoring, a change in ANC of at least 1–2 µeq/L per year over a period of about 10 years. The ability to detect a trend is also related to how long a monitoring program is continued. Small trends that are not detectable with 5 years of data can be obvious in 100 years. It requires a large trend magnitude to be detectable in a short amount of time. Trend detection is dependent on water quality parameter variability in terms of both analytical precision and natural (e.g., climatic) temporal variability. For a given trend magnitude of interest, it will take longer to detect a trend in a noisy, highly varying indicator than in a more precise and temporally stable indicator. Similarly, for a fixed amount of time to detect a trend, smaller trend magnitudes can be detected in stable indicators as compared with noisy indicators. For detecting regional trends (average trend across a number of sites), the number of regional sample sites is also an important factor. Regional trend detection ability increases with the number of sample sites. Thus, you can enhance your ability to document a trend by (1) monitoring over a longer time period, (2) reducing short-term variability caused by seasonality, episodes, or data quality, or (3) monitoring more sites. Larger trends will be easier to document than smaller trends. These issues need to be considered before you embark on a monitoring program intended to identify and quantify changes in water quality over time.

## 5.6   SUMMARY AND CONCLUSIONS

In summary, it is not possible to specify a routine set of data analyses that should be conducted for every water chemistry data set that might be assembled for the study of the effects of atmospherically deposited substances on water quality. Decisions regarding how to analyze the data will be influenced by the distribution of the collected data and the objectives of the particular investigation. Nevertheless, we suggest, using examples, many of the kinds of analyses that should be considered for the datasets commonly collected for this type of study.

The first step in analyzing a water chemistry dataset is to evaluate the overall quality of the data. In the process of evaluating data quality, it is often possible to identify data that are incorrect and sometimes to reanalyze or otherwise correct the identified errors. Next is a series of steps to prepare the raw data for graphical and statistical analysis. This involves applying procedures to deal with such potentially confounding issues as censored data, outliers, missing values, multiple observations, and treatment of zeros and negative values. Recommendations are provided here regarding how to deal with these issues.

A range of kinds of exploratory data analyses are illustrated, with examples from the published literature. These include suggestions regarding how to create subsets of the data to increase data analysis efficiency. Specific analyses suited to various study objectives are provided as examples.

Finally, the role of formal statistical analysis is considered. Frequently, such analyses will require the assistance of an individual who has formal training in statistics. In many cases, however, formal statistical analyses are not required. Much can be gained via routine exploratory analyses and application of simple graphical and analytical procedures.

# REFERENCES

Altman, D., T. Bryant, M. Gardner, and D. Machin. 2000. *Statistics with Confidence*. BMJ Books, London.

Baldigo, B.P., G.B. Lawrence, R.W. Bode, H.A. Simonin, K.M. Roy, and A.J. Smith. 2009. Impacts of acidification on macroinvertebrate communities in streams of the western Adirondack Mountains, New York, USA. *Ecol. Indicat.* 9:226–239.

Baldigo, B.P., G.B. Lawrence, and H.A. Simonin. 2007. Persistent mortality of brook trout in episodically acidified streams of the southwestern Adirondack Mountains, New York. *Trans. Am. Fish. Soc.* 136:121–134.

Charles, D.F., M.W. Binford, E.T. Furlong, R.A. Hites, M.J. Mitchell, S.A. Norton, F. Oldfield, M.J. Paterson, J.P. Smol, A.J. Uutala, J.R. White, D.F. Whitehead, and R.J. Wise. 1990. Paleoecological investigation of recent lake acidification in the Adirondack Mountains, NY. *J. Paleolimnol.* 3:195–241.

Charles, D.F. and J.P. Smol. 1990. The PIRLA II project: regional assessment of lake acidification trends. *Verh. Int. Ver. Theor. Angew. Limnol.* 24:474–480.

Cosby, B.J., R.F. Wright, G.M. Hornberger, and J.N. Galloway. 1985. Modelling the effects of acid deposition: estimation of long-term water quality responses in a small forested catchment. *Water Resour. Res.* 21(11):1591–1601.

Dise, N.B. and R.F. Wright. 1995. Nitrogen leaching from European forests in relation to nitrogen deposition. *For. Ecol. Manage.* 71:153–161.

Elser, J.J., T. Andersen, J.S. Baron, A.-K. Bergström, M. Jansson, M. Kyle, K.R. Nydick, L. Steger, and D.O. Hessen. 2009. Shifts in lake N:P stoichiometry and nutrient limitation driven by atmospheric nitrogen deposition. *Science* 326:835–837.

Gbondo-Tugbawa, S.S., C.T. Driscoll, J.D. Aber, and G.E. Likens. 2001. Evaluation of an integrated biogeochemical model (PnET-BGC) at a northern hardwood forest ecosystem. *Water Resour. Res.* 37(4):1057–1070.

Helsel, D.R. and R.M. Hirsch. 1992. *Statistical Methods in Water Resources.* Studies in Environmental Science 49. Elsevier, New York.

Hillman, D.C., S.H. Pia, and S.J. Simon. 1987. *National Surface Water Survey: Stream Survey (Pilot, Middle Atlantic Phase I, Southeast Screening, and Episode Pilot) Analytical Methods Manual.* EPA 600/8-87-005. US Environmental Protection Agency, Las Vegas, NV.

Hirsch, R.M. and J.R. Slack. 1984. A nonparametric trend test for seasonal data with serial dependence. *Water Resour. Res.* 20:727.

Irwin, R.J. 2008. *Checklist for Aquatic Vital Sign Monitoring Protocols and SOPs.* National Park Service, Water Resources Division, Fort Collins, CO.

Jeziorski, A., N.D. Yan, A.M. Paterson, A.M. DeSellas, M.A. Turner, D.S. Jeffries, B. Keller, R.C. Weeber, B.K. McNicol, M.E. Palmer, K. McIver, K. Arseneau, B.K. Ginn, B.F. Cumming, and J.P. Smol. 2008. The widespread threat of calcium decline in fresh waters. *Science* 322:1374–1377.

Lawrence, G.B., J.W. Sutherland, C.W. Boylen, S.A. Nierzwicki-Bauer, B. Momen, B.P. Baldigo, and H.A. Simonin. 2007. Acid rain effects on aluminum mobilization clarified by inclusion of strong organic acids. *Environ. Sci. Technol.* 41(1):93–98.

Lawrence, G.B., B.P. Baldigo, K.M. Roy, H.A. Simonin, R.W. Bode, S.I. Passy, and S.B. Capone. 2008a. *Results from the 2003–2005 Western Adirondack Stream Survey.* Final Report 08-22. New York State Energy Research and Development Authority (NYSERDA), Albany.

Lawrence, G.B., K.M. Roy, B.P. Baldigo, H.A. Simonin, S.B. Capone, J.W. Sutherland, S.A. Nierzwicki-Bauer, and C.W. Boylen. 2008b. Chronic and episodic acidification of Adirondack streams from acid rain in 2003–2005. *J. Environ. Qual.* 37:2264–2274. doi:2210.2134/jeq2008.0061.

Loftis, J.C., R.C. Ward, R.D. Phillips, and D.H. Taylor. 1989. *An Evaluation of Trend Detection Techniques for Use in Water Quality Monitoring Programs.* EPA/600/3-89/037. US Environmental Protection Agency, Environmental Research Laboratory, Corvallis, OR.

Long, R.P., S.B. Horsley, R.A. Hallett, and S.W. Bailey. 2009. Sugar maple growth in relation to nutrition and stress in the northeastern United States. *Ecol. Appl.* 19(6):1454–1466.

Minocha, R., W.C. Shortle, G.B. Lawrence, M.B. David, and S.C. Minocha. 1997. Relationships among foliar chemistry, foliar polyamines, and soil chemistry in red spruce trees growing across the northeastern United States. *Plant Soil* 191(1):109–122.

Moldan, F. and R.F. Wright. 1998. Episodic behavior of nitrate in runoff during six years of nitrogen addition to the NITREX catchment at Gårdsjön, Sweden. *Environ. Pollut.* 102(S1):439–444.

Paulsen, S. 1997. *Environmental Monitoring and Assessment Program: Integrated Quality Assurance Project Plan for Surface Waters Research Activities*. US Environmental Protection Agency, Office of Research and Development, National Health and Environmental Effects Research Laboratory, Corvallis, OR.

Rorabacher, D.B. 1991. Statistical treatment for rejection of deviant values: critical values of Dixon Q parameter and related subrange ratios at the 95 percent confidence level. *Anal. Chem.* 63(2):139–146.

SAS Institute. 1988. *SAS User's Guide: Statistics*. Version 5 edition. SAS Institute, Cary, NC.

Sen, P.K. 1968. On a class of aligned rank order tests in two-way layouts. *Ann. Math. Stat.* 39:1107–1380.

Stoddard, J.L. 1994. Long-term changes in watershed retention of nitrogen: its causes and aquatic consequences. In L.A. Baker (Ed.), *Environmental Chemistry of Lakes and Reservoirs*. American Chemical Society, Washington, DC, pp. 223–284.

Stoddard, J.L. 1995. Episodic acidification during snowmelt of high elevation lakes in the Sierra Nevada mountains of California. *Water Air Soil Pollut.* 85:353–358.

Stoddard, J., J.S. Kahl, F.A. Deviney, D.R. DeWalle, C.T. Driscoll, A.T. Herlihy, J.H. Kellogg, P.S. Murdoch, J.R. Webb, and K.E. Webster. 2003. *Response of Surface Water Chemistry to the Clean Air Act Amendments of 1990*. EPA 620/R-03/001. US Environmental Protection Agency, Office of Research and Development, National Health and Environmental Effects Research Laboratory, Research Triangle Park, NC.

Sullivan, T.J. 2000. *Aquatic Effects of Acidic Deposition*. Lewis/CRC Press, Boca Raton, FL.

Sullivan, T.J., B.J. Cosby, J.A. Laurence, R.L. Dennis, K. Savig, J.R. Webb, A.J. Bulger, M. Scruggs, C. Gordon, J. Ray, H. Lee, W.E. Hogsett, H. Wayne, D. Miller, and J.S. Kern. 2003. *Assessment of Air Quality and Related Values in Shenandoah National Park*. NPS/NERCHAL/NRTR-03/090. US Department of the Interior, National Park Service, Northeast Region, Philadelphia. http://www.nps.gov/nero/science/FINAL/shen_air_quality/shen_airquality.html.

Sullivan, T.J., B.J. Cosby, J.R. Webb, R.L. Dennis, A.J. Bulger, and F.A. Deviney Jr. 2008. Streamwater acid-base chemistry and critical loads of atmospheric sulfur deposition in Shenandoah National Park, Virginia. *Environ. Monitor. Assess.* 137:85–99.

Sullivan, T.J., J.M. Eilers, B.J. Cosby, and K.B. Vaché. 1997. Increasing role of nitrogen in the acidification of surface waters in the Adirondack Mountains, New York. *Water Air Soil Pollut.* 95(1–4):313–336.

Sullivan, T.J., D.L. Kugler, M.J. Small, C.B. Johnson, D.H. Landers, B.J. Rosenbaum, W.S. Overton, W.A. Krester, and J. Gallagher. 1990. Variation in Adirondack, New York, lakewater chemistry as a function of surface area. *Water Resour. Bull.* 26:167–176.

Sullivan, T.J., J.R. Webb, K.U. Snyder, A.T. Herlihy, and B.J. Cosby. 2007. Spatial distribution of acid-sensitive and acid-impacted streams in relation to watershed features in the southern Appalachian mountains. *Water Air Soil Pollut.* 182:57–71.

# Chapter 6

# Field Sampling for Aquatic Biota

## 6.1 BACKGROUND

Aquatic invertebrates can be good indicators of water quality and can provide documentation of ecological effects of changing water quality. Bottom-dwelling (benthic) invertebrates have been used extensively to assess biological conditions in streams. Benthic macroinvertebrates can also be used in assessing lake biology, but their use for this purpose has not been common in the United States. More commonly, in this country, biological conditions in the epilimnion of thermally stratified lakes are evaluated using zooplankton. Both stream macroinvertebrate and lake zooplankton data can provide useful information to reveal some of the ecological effects that result from atmospheric deposition of acid precursors, nutrients, or toxic substances and consequent alterations of surface water chemistry.

### 6.1.1 Lake Zooplankton

Lake zooplankton include crustaceans, rotifers, pelagic insect larvae, and aquatic mites. Many species, especially of the crustaceans and rotifers, are known to be sensitive to changes in water chemistry (cf. Melack et al. 1989, Gerritsen et al. 1998, Sullivan et al. 2006). Nevertheless, the species composition and trophic structure of zooplankton communities are controlled by multiple factors, of which aquatic acid-base chemistry, nutrient availability and stoichiometry, and concentrations of toxins are only a few possibilities. Populations of zooplankton can be strongly influenced by changes at both lower and higher trophic levels because zooplankton are sensitive to changes in the distribution and abundance of both algae and predators. Predation occurs by planktonic predators and by fish. Thus, the presence or absence of plankton-feeding fish can have a large influence on the presence, abundance, and body size of various species of zooplankton. It can therefore be difficult to infer the causes of observed changes in the zooplankton community over time unless data are also available regarding the status of the fish populations in the lakes under study. Therefore, if fish population data are available for lakes within a given forest or wilderness, it can be helpful to select lakes for zooplankton sampling that also have data on fish. In addition, both intra-annual and interannual variability in zooplankton species distributions can be high. In general, zooplankton are at their greatest development from June to mid-October, and midsummer is considered to be a relatively stable period for zooplankton monitoring.

Zooplankton are important components of the biological community of lakes. There may be as many as 200 species or more that occur within lakes in a given region. They constitute key portions of the aquatic food web and play a major role in transferring energy from the primary producers (mainly phytoplankton) to predatory invertebrates and to fish and other vertebrates. Individual zooplankton species, and the zooplankton community as a whole, respond to a number of environmental stressors. These include acidification, nutrient enrichment, sedimentation, fish stocking, and habitat manipulation. Effects of these environmental stressors can sometimes be revealed by evaluating changes in the presence/absence of known regional indicator species, overall species composition, biomass, body size distribution, or the structure of the food web.

## 6.1.2   Stream Macroinvertebrates

Benthic macroinvertebrates inhabit the bottom substrates of streams and provide a good indication of overall biological condition (Kerans and Karr 1994, Barbour et al. 1999, Reynoldson et al. 2001, Klemm et al. 2002, 2003, Clarke et al. 2003, Bailey et al. 2004, Griffith et al. 2005). Monitoring these assemblages is useful in assessing the status of the water body and investigating the possibility of trends over time in ecological condition. Benthic macroinvertebrate species respond to a variety of stressors in different ways, and it is often possible to determine the type of stress that has affected the macroinvertebrate assemblage (e.g., Klemm et al. 2002). Because many stream macroinvertebrates have life cycles of a year or more and are relatively immobile, macroinvertebrate assemblage structure is a function of present and past conditions and provides an integration of the variability that typically occurs in stream condition with season and with changing hydrology (Barbour et al. 1999, Peck et al. 2006). For general bioassessment purposes, stream macroinvertebrates are typically sampled in summer. However, for assessing acidic deposition impacts, we generally recommend sampling during spring base flow as that is the season of maximum impact (lowest acid-neutralizing capacity [ANC]/pH).

At high elevation in regions that experience substantial snowpack development, spring sampling may not be feasible; at such locations, summer sampling is recommended.

The insect order Ephemeroptera (mayflies) is an excellent indicator taxa for acidification effects. However, it comprises just one order of benthic invertebrates, and there are many other taxonomic groups that make up the benthic stream community that would need to be taken into account for a full biological assessment. These other taxa also contribute a great deal of information about stream condition. In addition to acidification impacts, macroinvertebrates are excellent indicators of substrate alteration (e.g., sedimentation), nutrient enrichment, metal pollution, and habitat alteration. In the Environmental Protection Agency (EPA) national wadeable stream assessment, the greatest risk of having poor stream macroinvertebrate condition was found in streams with excess sediment (enhanced erosion) or nutrient enrichment (Van Sickle and Paulsen 2008).

Some studies have found that acidified streams host fewer invertebrate taxa than streams with higher ANC and pH (e.g., Feldman and Connor 1992, Kauffman et al. 1999, Sullivan et al. 2003). This is especially true for mayflies and to a lesser extent for caddis flies (order Tricoptera) and stone flies (order Plecoptera). Aquatic insect status is sometimes evaluated on the basis of these three orders using what is known as the EPT (Ephemeroptera-Plecoptera-Tricoptera) Index (EPT taxa richness).

## 6.2   AQUATIC INVERTEBRATE STUDY DESIGN

Aquatic invertebrates can be collected and analyzed as part of lake or stream characterization studies, synoptic surveys, or long-term monitoring or be used to augment model projections of future chemical conditions. As is described in detail in Chapter 2, the design of an aquatic biota study should be a function of the study purpose and questions asked. Some example approaches for biological characterization or monitoring are outlined in Table 6.1.

Each example approach is tied to a specific purpose. The reader is also referred to the discussion with examples provided in Chapter 1.

Evaluation of the status of the aquatic invertebrate biota can be used, along with assessment of chemical status or change in water quality conditions, to estimate the impacts of nutrient enrichment, acidification, and various kinds of habitat disturbances. In general, changes in aquatic chemistry are more easily documented than are changes in aquatic biology. Nevertheless, the concerns on the part of land managers and the public regarding changes in chemistry are fundamentally rooted in widespread concerns about protecting

**TABLE 6.1    EXAMPLE APPROACHES FOR BIOLOGICAL SAMPLING TIED TO THE PURPOSE OF THE FIELD STUDY**

| Purpose | Approach |
|---|---|
| 1. Determine spatial patterns in biological assemblages relative to chemical or deposition gradients | a. Conduct survey of lake zooplankton or stream benthic macroinvertebrate communities in waters that exhibit varying water chemistry or that receive varying atmospheric deposition levels |
| 2. Quantify long-term changes over time in biology | a. Conduct trends analysis in species richness<br>b. Compare trends in biota with trends in water chemistry |
| 3. Determine extent to which air pollution is affecting water resources | a. Characterize biology of multiple lakes or streams expected to be sensitive<br>b. Plot changes in species richness versus changes in ANC or $NO_3^-$ concentration |

resources against biological damage. Thus, the ultimate purpose of studying or monitoring aquatic chemistry is often mainly to aid in the protection of biological resources. Also, biota reflect conditions over longer time periods than the single point in time represented by a water sample. Therefore, there is additional power in the inclusion of a biological component in the investigation or monitoring of chemical conditions.

For evaluation of biological responses to acidification, study designs most commonly include (1) documentation across sites (lakes or stream reaches) within a reasonably small area (i.e., one wilderness area) of relationships between water chemistry (usually ANC; can also include pH, inorganic monomeric aluminum [$Al_i$] or nitrate [$NO_3^-$] concentration) and macroinvertebrate taxonomic composition or (2) evaluation of changes over time in water chemistry and macroinvertebrate taxonomic composition. Such studies often focus on aquatic insects in the orders that include mayflies, caddis flies, and stone flies for streams. For a more rapid and less-expensive stream assessment, the analysis can be restricted to only mayflies, which is the order most susceptible to acidity. However, there may be no mayfly species present at all if the ANC in a given study stream is especially low (near or below zero). For lakes, acidification studies typically focus on zooplankton, mainly crustaceans and rotifers. We recommend, if available funding allows, that such studies be conducted on a suite of acid-sensitive lakes or streams within a given study area or region if lake or stream acidification is believed to be an important issue. If, after a prolonged period of monitoring, a trend is indeed documented, a decision will need to be made regarding whether to continue monitoring. In general, we recommend continued monitoring for the foreseeable future. These will likely be important data that will help in the future to sort out interactions between climate change and air pollution effects.

For a more complete assessment of biological condition, we recommend use of an Index of Biotic Integrity (IBI). Such an index provides a more complete assessment of biological condition than the rapid single-order or EPT assessments discussed previously. It does require compiling available information about the feeding groups, pollution tolerance, and habits of the taxa in the study waters to calculate the IBI. For this reason, we do not necessarily recommend implementation of an IBI as a routine procedure. It can provide a more rigorous assessment of biological conditions and response to multiple stressors where in-depth study is warranted. It is somewhat more expensive, however, and requires more specialized taxonomic and autecological (individual species ecology) expertise.

## 6.3   SITE SELECTION

In general, criteria for site selection for the purpose of conducting a biological assessment are the same as criteria for site selection for the purpose of investigating aquatic chemistry. The issues of sampling site location, random versus nonrandom site selection, and the establishment and documentation of the stream reaches or lakes to sample are the same for biological studies as for chemical studies. See Section 2.2.1 for more complete discussion of these issues.

By necessity in many cases, the sites included in a biological study will be only a subset of the sites included in the chemical investigation. It is important fiscally to choose this subset wisely. In general, one may wish to avoid sites having substantial disturbance other than atmospheric deposition (i.e., geological sulfur, forest fire, insect infestation, tree disease, large windthrow, or other substantial disturbance) that may influence the acid-base chemistry or nutrient status of drainage waters. In general, one should include for biological characterization or monitoring sites that are expected to be sensitive to the stresses of interest. For acidification studies, these are usually the lakes or streams having ANC less than about 50 to 100 µeq/L. For nutrient enrichment studies, these are usually N-limited water bodies. Short of direct experimentation, it is difficult to predict which water bodies might be N limited. The Redfield ratio, based on the molar N:P ratio in phytoplankton, suggested that water bodies might be N limited if the molar N:P ratio was less than 16. Subsequently, experimental studies suggested higher cutoffs. Recent research suggests that N-limited lakes generally include lakes having the molar ratio of N:P less than about 44 (Guildford and Hecky 2000, Schindler et al. 2008, Elser et al. 2009).

## 6.3.1    Laying Out the Support Reach for Stream Macroinvertebrate Sampling

Unlike chemistry, which can be measured at one point, characterizing stream biota usually requires sampling a length of a stream that captures the range of available habitat. There are a large number of field protocols for sampling stream macroinvertebrates for bioassessment. They all specify collecting a number of different net samples (kick, Hess, or Surber) from different places along the stream sample reach and compositing them into either a single composite sample or habitat type (e.g., riffle) composite sample. The procedures that we recommend (summarized here) are based on the procedures developed by the US EPA for the Environmental Monitoring and Assessment Program Surface Water (EMAP-SW) sampling program (Peck et al. 2006). These protocols have been used in studies of streams across the entire United States, so they work in a wide variety of stream types. They were also designed to be implemented by many different field crews and require minimal in-field decision making.

At each selected stream-sampling location (called the X site), the support reach (the length of stream to be sampled at the sampling location) must be laid out. The support reach must be sufficiently long to represent the biological community being sampled. Based on several studies (Robison 1998, Li et al. 2001, Reynolds et al. 2003), a support reach with a length of 40 times the average wetted channel width measured near the X site is sufficient for almost all sites. The support reach is established about the X site using the procedures described in Table 6.2. Field staff should reconnoiter the support reach to make sure it is clear of obstacles that would prohibit sampling and data collection activities. Record the channel width used to determine the support reach length and identify the support reach length upstream and downstream of the sample site. Figure 6.1 illustrates the principal features of a hypothetical support reach, including the location of 11 cross-sectional transects from which the samples will be collected.

There are some conditions that may require adjusting the support reach about the X site (i.e., the X site will be shifted either upstream or downstream and will therefore not be located at the midpoint of the support reach) to avoid features we do not wish to (or physically cannot) sample across. The full length of the support reach should be of the same stream order as the X site. Do not proceed upstream into a stream reach if the stream order decreases or downstream into a stream reach if the stream order increases. If you encounter an impoundment, such as a lake, reservoir, or pond, or an impassible barrier (e.g., a waterfall, a cliff) while laying out the support reach, adjust the reach such that the barrier is at one end. Adjusting, or *sliding*, the support reach involves noting the distance of the confluence, barrier, or other restriction from the X site; flagging the confluence, impoundment, or barrier as the end point of the reach; and adding the distance to the other end of the reach, such that

**TABLE 6.2   LAYING OUT THE SUPPORT REACH**

A. Use a surveyor's rod or tape measure to determine the wetted width of the channel at 3 to 5 places considered to be of *typical wetted width* within approximately 5 channel widths upstream and downstream from the X site. Average the readings together and round to the nearest 0.5 m. If the average width is less than 3.5 m, use 150 m as a minimum support reach length. Record this width on the Stream Verification Form.

   For channels with *interrupted flow*, estimate the width based on the unvegetated width of the channel (again, with a 150-m minimum).

B. Check the condition of the stream upstream and downstream of the X site by having one team member go upstream and one downstream. Each person goes until he or she has visited the candidate sample reach to a distance of 20 times the average channel width in each direction (equal to one-half the support reach length, but a minimum of 75 m) determined in step 1 from the X site.

   For example, if the support reach length is determined to be 150 m, each person would go 75 m from the X site to lay out the reach boundaries.

C. Determine if the support reach needs to be adjusted about the X site because of confluences with higher-order streams (downstream); lower-order streams (upstream); impoundments (lakes, reservoirs, ponds); or physical barriers (e.g., falls, cliffs) or because of access restrictions to a portion of the initially determined support reach.

   If such a confluence, barrier, or access restriction is present, note the distance and flag the confluence, barrier, or the limit of access as the endpoint of the reach. Move the other endpoint of the support reach an equivalent distance away from the X site. *The X site must still be* **within the support reach** *after adjustment*. The total support reach length does not change, but the support reach is no longer centered on the X site.

   **Note:** If the sampling sites are statistically (randomly) selected, **do not** slide the support reach to avoid man-made obstacles such as bridges, culverts, riprap, or channelization or in streams with interrupted flow to obtain more inundated areas to sample. If the sites are not statistically selected, it is recommended to avoid sites that are influenced by substantial human-caused channel disturbance.

D. Starting back at the X site (or the new midpoint of the reach if it had to be adjusted as described in step C), measure a distance of 20 channel widths down one side of the stream using a tape measure. Be careful not to "cut corners." Enter the channel to make measurements only when necessary to avoid disturbing the stream channel prior to sampling activities. This end point is the downstream end of the support reach and is flagged as the location of transect A.

*Continued*

**TABLE 6.2 (*Continued*)    LAYING OUT THE SUPPORT REACH**

E.  Using the tape measure, measure 1/10 (4 channel widths in big streams or 15 m in small streams) of the required stream length upstream from the start point (transect A). Flag this spot as the next cross section or transect (transect B). For transect A, roll one die to determine if it is a left (L), center (C), or right (R) sampling point ( following the convention of facing downstream) for collecting benthic macroinvertebrate samples. A dice roll of 1 or 2 indicates L, 3 or 4 indicates C, and 5 or 6 indicates R (or use a digital wristwatch and glance at the last digit (1 – 3 = L, 4 – 6 = C, 7 – 9 = R). Mark L, C, or R on the transect flagging.

F.  Proceed upstream with the tape measure and flag the positions of 9 additional transects (labeled C through K as you move upstream) at intervals equal to 1/10 of the reach length. Assign sampling spots to each transect in order as L, C, R after the first random selection.

For example, if the sampling spot assigned to transect A was *C*, transect B is assigned *R*, transect C is *L*, transect D is *C*, and so on.

the total support reach length remains the same, but it is no longer centered about the X site. If you are denied access permission to a portion of the support reach, you can adjust the reach to make it entirely accessible; use the point of access restriction as the end point of the reach.

## 6.3.2   Lake Selection for Zooplankton Sampling

Acidification and eutrophication effects on individual species of zooplankton and on the zooplankton community in general may occur across a rather wide spectrum of lake pH, ANC, $Al_i$, and nutrient concentrations. For acidification, effects are usually observable at ANC values below about 50 to 100 µeq/L and pH below about 6.0 to 6.5. Such ANC and pH cutoff values generally correspond with $Al_i$ concentrations near 2 µM. Lakes having pH below 6.0 or ANC below about 50 to 100 µeq/L have an increased likelihood of having $Al_i$ above this general response threshold. Nevertheless, it is possible that effects on zooplankton also occur at somewhat higher pH and ANC (and lower $Al_i$) values. A complicating factor relates to the influence of lake size and watershed area on lake biology. In general, smaller lakes in smaller watersheds are more likely to be lower in pH and ANC and to have less-diverse zooplankton communities than larger lakes in larger watersheds. Certainly, to some degree (often to a large degree) this relationship is controlled by the effects of lake chemistry on biota. But also, smaller lakes in smaller watersheds might be expected to have less-diverse biotic assemblages than larger lakes in larger watersheds as a consequence of their physical simplicity and reduced number of available niches. Thus, the often-observed patterns of changing zooplankton species composition and taxonomic richness among lakes are likely only partly caused by water acid-base chemistry. This makes it difficult to tease

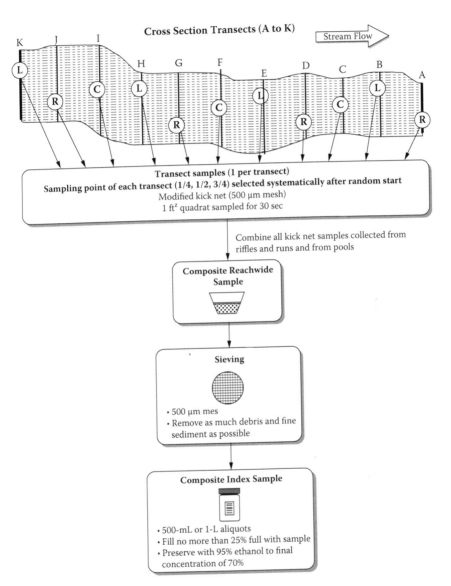

**Figure 6.1**   Design for the reachwide benthic macroinvertebrate sample. (From Peck et al. 2006. *Environmental Monitoring and Assessment Program-Surface Waters Western Pilot Study: Field Operations Manual for Wadeable Streams.* EPA/620/R-06/003. US Environmental Protection Agency, Office of Research and Development, Washington, DC.)

out the effects of changing water chemistry when evaluating changes over time in the biological community. For these (and other) reasons, it is important to give careful consideration to site selection for zooplankton monitoring or characterization. It can be helpful to include multiple lakes, selected to cover a range of acid-base chemistry and perhaps within rather narrow windows regarding lake and watershed areas. In addition, it can be helpful to study the zooplankton communities of lakes that are also being studied with regard to their fish and algal communities. This may allow an improved opportunity to sort out what may be a multitude of factors that simultaneously influence the lake zooplankton community. Biological effects are more likely to be observable in lakes that have relatively low ANC (<50 µeq/L) and pH (<6.0). Nevertheless, having lakes in the study with somewhat higher ANC and pH is also important, especially for evaluating effects on taxonomic richness or the presence/absence of particular indicator species. This will help to make sure that a sufficient range of response occurs in the dataset to increase the likelihood of being able to document what may be a noisy relationship. Further discussion of the interpretation of zooplankton data is provided in Section 6.9. It is likely that study of the zooplankton community will be less helpful for evaluation of effects related to nutrient N enrichment or the deposition of toxic substances. Such effects on zooplankton communities have not been as well documented as the effects from acidification.

Zooplankton samples should generally be collected at the lake index location for water chemistry sampling. This should be the deepest part of the lake. The index location is described in the chapter on water chemistry sampling. It is important to collect the zooplankton tows in the deepest part of the lake because, especially in mid- to late summer, the size of the cold-water hypolimnion can be reduced substantially. Missing the deep spot can cause exclusion of individuals occupying the cold-water stratum, thereby confounding interpretation of the true zooplankton assemblage in the lake.

Zooplankton tows can be compromised by high algal production or high dissolved organic carbon (DOC) because of algal or organic particle fouling of the net. This problem can be partially ameliorated by using a reducing collar attached to the plankton net. Alternatively, if fouling is a major problem, zooplankton can be collected as integrated water column samples using a hose-and-pump system.

## 6.4   PRETRIP PREPARATION

### 6.4.1   Equipment and Supplies

Table 6.3 shows the checklist of equipment and supplies required to complete the collection of benthic macroinvertebrates from streams. Use this checklist to ensure that equipment and supplies are organized and available at the stream

**TABLE 6.3 EQUIPMENT AND SUPPLIES FOR BENTHIC MACROINVERTEBRATES**

| Quantity | Item | √[a] |
|---|---|---|
| 1 | Modified kick net (D frame with 500-μm mesh) and 4-ft handle | |
| | Spare net(s) or spare bucket assembly for end of net | |
| 1 | Watch with timer or a stopwatch | |
| 2 | Buckets, plastic, 8- to 10-qt capacity (collapsible for backcountry) | |
| 1 | Sieve with 500-μm mesh openings or sieve-bottomed bucket, 500-μm mesh openings | |
| 2 pair | Watchmaker's forceps (straight and curved) | |
| 1 | Wash bottle, 1-L capacity, labeled *STREAM WATER* | |
| 1 | Small spatula, spoon, or scoop to transfer sample | |
| 1 | Funnel, with large-bore spout | |
| 4 to 6 each | Sample jars, HDPE plastic with leakproof screw caps, 500-ml or 1-L capacity, suitable for use with ethanol | |
| 2 gal | 95% ethanol, in a proper container (smaller amounts can be carried in for backcountry work or ethanol can be added at the vehicle after returning from the field) | |
| 2 pair | Rubber gloves suitable for use with ethanol | |
| 1 | Cooler (with suitable absorbent material) for transporting ethanol and samples in vehicle | |
| 2 | Preprinted benthic sample labels with sample ID numbers | |
| 4 | Preprinted benthic sample labels without sample ID numbers | |
| 6 | Blank labels on waterproof paper for inside jars | |
| 1 | Sample Collection Form for site | |
| | Soft (no. 2) pencils | |
| | Fine-tip indelible markers | |
| 1 package | Clear tape strips | |
| 4 rolls | Plastic electrical tape | |
| 1 | Pocketknife, with at least two blades | |
| 1 | Scissors | |
| 1 | Pocket-size field notebook (optional) | |
| 1 package | Kim wipes in small resealable plastic bag | |

*Continued*

**TABLE 6.3 (*Continued*)    EQUIPMENT AND SUPPLIES FOR BENTHIC MACROINVERTEBRATES**

| Quantity | Item | $\checkmark$[a] |
|---|---|---|
| 1 set | Laminated sheets of procedure tables or quick reference guides for benthic macroinvertebrates | |

*Source:* Peck et al. 2006. *Environmental Monitoring and Assessment Program-Surface Waters Western Pilot Study: Field Operations Manual for Wadeable Streams.* EPA/620/R-06/003. US Environmental Protection Agency, Office of Research and Development, Washington, DC.

*Note:* HDPE = high-density polyethylene.

[a] Check off each item as it is added.

site so the activities can be conducted efficiently. Similarly, Table 6.4 provides the checklist for zooplankton sampling.

## 6.4.2    Equipment-Cleaning Protocols

Field survey personnel or their equipment can serve to transport pathogens and invasive species among water bodies. Field personnel should take appropriate precautions to minimize or eliminate this risk. General equipment-cleaning guidelines are provided here. In addition, field staff should consult with local experts to determine if local conditions require any additional special precautions.

Between sample sites and at the duty station subsequent to field sampling, all gear that was exposed to stream or lake water should be thoroughly cleaned. Clothing, skin, and fingernails should also be cleaned. Gear should be disinfected using a 10% bleach solution, or a solution of an alternative disinfectant product, and thoroughly rinsed. Use of a high-pressure hose can be helpful. Gear should then be completely dried prior to reuse at another lake or stream.

Note that it is important to follow appropriate safety precautions when working with disinfectant products, especially the concentrated solutions. Such precautions include appropriate ventilation, use of impervious gloves and splash goggles, and access to eyewash stations.

## 6.5    COLLECTION PROCEDURES

### 6.5.1    Stream Benthic Macroinvertebrates

The procedures recommended in this protocol for collection and preservation of stream macroinvertebrates are based largely on the protocol designed by the EPA for the EMAP-SW surface-water-sampling efforts. EMAP-SW protocols for invertebrate sampling are described in detail by Peck et al. (2006).

**TABLE 6.4 EQUIPMENT AND SUPPLIES FOR COLLECTING ZOOPLANKTON SAMPLES**

| Quantity[a] | Item | √[b] |
|---|---|---|
| 2 | Wisconsin fine-mesh (80-μm) net[c] with attached collection bucket | |
| 2 | Wisconsin coarse-mesh (243-μm) net[c] with attached collection bucket | |
| 2 | Sample line, marked at 0.5-m increments | |
| 2 | Secchi disk with cable | |
| 2+/site | 125-ml wide-mouth polyethylene sample jars (two per site, plus additional for replicates and other backup sampling) | |
| 1 | Squirt bottle with deionized water (DIW) | |
| | 95% ethanol | |
| 2+/site | $CO_2$ tablets | |
| 1+/site | 500-ml wide-mouth container | |
| 2 | Two lids converted to form strainers (one with 80-μm, one with 243-μm mesh), made by drilling two holes in each lid and gluing a piece of the netting to the inside of the lid using silicone glue | |
| 2+/site | Ziplock-type plastic bag | |
| | Clear tape for covering labels | |
| | Electrical tape | |
| 1+/site | Zooplankton Sample Data Form | |
| | Pencils and permanent markers | |
| | Mild (10%) bleach solution for cleaning net and strainer lids between lakes; backwash net with a garden hose after use | |

*Source:* Modified from US Environmental Protection Agency. 2007. *Survey of the Nation's Lakes. Field Operations Manual.* EPA 841-B-070004. US Environmental Protection Agency, Washington, DC.

a  It is advisable to include some extras beyond what is needed for the number of sites to be sampled.
b  Check off each item as it is added.
c  These two mesh sizes (80 and 243 μm) are general guidelines. Other sizes could be used.

The EMAP-SW benthic macroinvertebrate protocol was designed to evaluate the biological condition of wadeable streams in the United States for the purpose of detecting stresses on assemblage structure and assessing the relative severity of these stresses (Peck et al. 2006). It is based on the level III procedure for benthic macroinvertebrates of the EPA Rapid Bioassessment

Protocol (RBP; Plafkin et al. 1989, Barbour et al. 1999), which has been adopted for use by many states.

Benthic macroinvertebrates are collected at each of 11 equidistant transects spaced throughout the support reach to ensure distribution of individuals among available major habitat types, eliminate individual sampler bias, and provide a comparable and consistent sample from every reach. All 11 transect samples are combined into a single composite sample to characterize the support reach and reduce the cost and effort in processing and analysis (Patil et al. 1994, Barbour et al. 1999, Roth et al. 2002). The number of individual field collections (11) is expected to provide a composite sample with a sufficient number of individuals to characterize the taxonomic composition and relative abundance of the stream assemblage (e.g., Larsen and Herlihy 1998).

Samples are collected from each support reach with a D-frame kick net that can generally be used in the stream by one person (Figure 6.2). Typically, a field crew of two people collect kick net samples for benthic macroinvertebrates. One person will collect the samples while a second person times the collection of samples and records information on the field data form. However, in swift waters, two people may be needed to collect the samples.

Each kick net sample is collected at each of the 11 cross-sectional transects (transects A through K) at an assigned sampling point (left, center, or right) as illustrated in Figure 6.1. Assign the left, center, or right sampling point at transect A at random. Once the first sampling point is determined, assign points at successive transects in order (left, center, right). At transects assigned a center sampling point where the stream width is between one and two net widths wide, pick either the left or right sampling point instead. If the stream is only one net width wide at a transect, place the net across the entire stream width and consider the sampling point to be center. If a sampling point is located in water that is too deep or otherwise unsafe to wade, select an alternate sampling point nearby. Never sample at an unsafe location.

Collect a kick net sample at each transect as described in Table 6.5 beginning at the transect that is furthest downstream. Never collect benthic macroinvertebrates from a streambed location that you have recently disturbed (e.g., walked in). If a replicate composite sample is to be collected, do so at each transect within that support reach before moving upstream to the next transect. At each sampling point, determine if the flowing water or the slack water procedure is to be used, based on whether there is enough current to extend the net. These procedures are described in Table 6.5. For each kick net sample, record the dominant substrate type (fine/sand, gravel, coarse substrate [coarse gravel or larger] or other [e.g., bedrock, hardpan, wood, aquatic vegetation, etc.]) and the habitat type (pool, glide, riffle, or rapid) on the Sample Collection Form. Note that these substrate types and habitats are defined in the table. Collect only from the upper 4 to 5 cm (1.5 to 2 in.) of the substrate. As you go upstream from transect to transect,

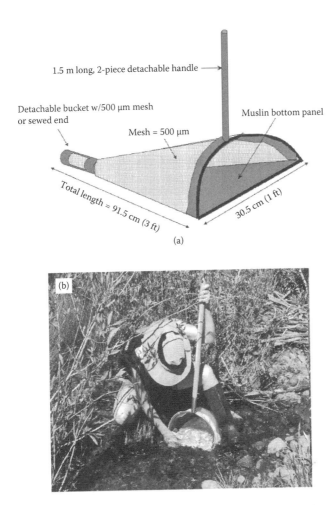

**Figure 6.2** Modified D-frame kick net: (a) schematic drawing (not drawn to scale) (From Peck et al. 2006. *Environmental Monitoring and Assessment Program-Surface Waters Western Pilot Study: Field Operations Manual for Wadeable Streams.* EPA/620/R-06/003. US Environmental Protection Agency, Office of Research and Development, Washington, DC); (b) photograph showing EMAP crew sampling macro-invertebrates with modified D-frame net in Utah (Photo by A. Herlihy).

combine all the kick net samples into a container, ignoring whether they were collected using the flowing water or slack water procedure.

If the kick net cannot be used to collect a sample at a flowing water sampling point, select the number of rocks necessary to cover approximately 1 ft$^2$ (0.09 m$^2$) of the streambed from the area near the sampling point (within the area of

**TABLE 6.5    PROCEDURE TO COLLECT KICK NET SAMPLES FOR THE REACHWIDE COMPOSITE SAMPLE**

1. At each cross-sectional transect, beginning at the downstream end of the reach with transect A (Figure 6.1), locate the assigned sampling point (left, center, or right as you face downstream) as 25%, 50%, and 75% of the wetted width, respectively. If you cannot collect a sample at the designated point because of deep water or unsafe conditions, relocate the point nearby on the same transect.

2. Attach the handle to the kick net. Make sure that the handle is attached tightly or the net may become twisted in a strong current, causing the loss of part of the sample.

3. Determine if there is sufficient current in the area at the sampling point to extend the net fully. If so, use the *flowing water* procedure (go to step 4). If not, use the *slack water* procedure (go to step 10).

   For vegetation-choked sampling points where neither procedure can be used, sweep the net through the vegetation within a 0.09-m² (1-ft²) quadrat for 30 s. Place the contents of this handpicked sample into the sampling container. Go to step 14.

### Flowing Water Procedure

4. With the net opening facing upstream, position the net quickly and securely on the stream bottom to eliminate gaps under the frame. Avoid large rocks that prevent the sampler from seating properly on the stream bottom.

   **Note:** If there is too little water to collect the sample with the kick net, randomly pick up 10 rocks from the riffle and pick and wash the organisms off them into a bucket that is half-full of water.

5. Holding the net in position on the substrate, visually define a rectangular quadrat that is one net width wide and one net width long upstream of the net opening. The area within this quadrat is about 0.09 m² (1 ft²). Alternatively, place a wire frame of the correct dimensions in front of the net to help delineate the quadrat to be sampled.

6. Hold the net in place with your knees. Check the quadrat for heavy organisms, such as mussels and snails. Endangered species must be left in place and not removed, but they need to be recorded on the field form. Remove any heavy nonendangered species from the substrate by hand and place them into the net. Pick up any loose rocks or other larger substrate particles in the quadrat. Use your hands or a small scrub brush to dislodge organisms so that they are washed into the net. Scrub all rocks that are golf ball size or larger and are situated over halfway into the quadrat. Large rocks that are less than halfway into the sampling area are pushed aside. After scrubbing, place the substrate particles outside the quadrat.

7. Keep holding the sampler securely in position. Start at the upstream end of the quadrat and use your foot and toes to *vigorously* kick the *upper 4 to 5 cm (1.5 to 2 in.)* of the remaining finer substrate within the quadrat for 30 s (use a stopwatch). Avoid kicking too deep into the substrate.

**TABLE 6.5 (*Continued*)   PROCEDURE TO COLLECT KICK NET SAMPLES FOR THE REACHWIDE COMPOSITE SAMPLE**

**Note:** For samples located within dense beds of long, filamentous aquatic vegetation (e.g., algae or moss), kicking within the quadrat may not be sufficient to dislodge organisms in the vegetation. Usually, these types of vegetation are lying flat against the substrate because of current. Use a knife or scissors to remove *only the vegetation that lies **within** the quadrat* (i.e., not entire strands that are rooted within the quadrat) and place it into the net.

8. Pull the net out of the water. Immerse the net in the stream several times to remove fine sediments and to concentrate organisms at the end of the net. Avoid having any water or material enter the mouth of the net during this operation.

9. Go to step 14.

**Slack Water Procedure**

10. Visually define a rectangular quadrat that is one net width wide and one net width long at the sampling point. The area within this quadrat is about 0.09 m² (1 ft²). Alternatively, lay a wire frame of the correct dimensions in front of the net at the sampling point to help delineate the quadrat.

    **Note:** If there is not enough water present to use the net, spend 30 s collecting and examining pieces of substrate from about 0.09 m² (1 ft²) of substrate at the sampling point.

11. Inspect the stream bottom within the quadrat for any heavy organisms, such as mussels and snails. Remove these organisms by hand and place them into the net or into a bucket. Pick up any loose rocks or other larger substrate particles within the quadrat and hold them in front of the net. Use your hands (or a scrub brush) to rub any clinging organisms off rocks or other pieces of larger substrate (especially those covered with algae or other debris) into the net. After scrubbing, place the larger substrate particles outside the quadrat.

12. Use your foot and toes to *vigorously* kick the *upper 4 to 5 cm (1.5 to 2 in.)* of the remaining finer substrate within the quadrat while dragging the net repeatedly through the disturbed area just above the bottom. Keep moving the net all the time so that the organisms trapped in the net will not escape. Continue kicking the substrate and moving the net for 30 s.

    **Note:** If there is too little water to use the kick net, *vigorously* stir up the substrate with your gloved hands and use a sieve with 500-μm mesh to collect the organisms from the water in the same way the net is used in larger pools.

13. After 30 s, remove the net from the water with a quick upstream motion to wash the organisms to the bottom of the net.

*Continued*

**TABLE 6.5 (*Continued*)   PROCEDURE TO COLLECT KICK NET SAMPLES FOR THE REACHWIDE COMPOSITE SAMPLE**

### All samples

14. Invert the net and transfer the contents into a bucket or wide-mouth container with a lid marked *reachwide*. Inspect the net for any residual organisms clinging to the net and deposit them into the reachwide container. Use a squirt bottle with stream water and watchmaker's forceps if necessary to remove organisms from the net. Carefully inspect any large objects (such as rocks, sticks, and leaves) in the bucket and wash any organisms found off the object and into the bucket before discarding the object. Remove as much detritus as possible *without losing* any organisms. Replace the lid on the bucket or container.

15. Determine the *predominant* substrate size/type you observed within the sampling quadrat. Place an *X* in the appropriate substrate type box for the transect on the Benthic Macroinvertebrate Sample Collection Form.

    **Note:** If there are codominant substrate types, you may check more than one box; note the codominants in the comments section of the form.
      Fine/sand: not gritty (silt/clay/muck < 0.06-mm diameter) to gritty, up to ladybug-size (2-mm diameter)
      Gravel: fine-to-coarse gravel (ladybug to tennis ball size; 2- to 64-mm diameter)
      Coarse: Cobble to boulder (tennis ball to car size; 64 mm to 4000 mm)
      Other: bedrock (larger than car size; > 4000 mm), hardpan (firm, consolidated fine substrate), wood of any size, aquatic vegetation, and so on. Note type of "other" substrate in comments on field form.

16. Identify the habitat type where the sampling quadrat was located. Place an *X* in the appropriate channel habitat type box for the transect on the Sample Collection Form.
      Pool: still water; low velocity; with smooth, glassy surface; usually deep compared to other parts of the channel.
      GLide: water moving slowly, with smooth, unbroken surface; low turbulence.
      RIffle: water moving, with small ripples, waves, and eddies; waves not breaking, and surface tension is not broken; "babbling" or "gurgling" sound.
      RApid: Water movement is rapid and turbulent; surface with intermittent "white water" with breaking waves; continuous rushing sound.

17. Proceed upstream to the next transect (including all transects through transect K, the upstream end of the support reach) and repeat steps 1 through 16. Combine all kick net samples within the sample reach into the reachwide container.

**TABLE 6.5 (*Continued*)  PROCEDURE TO COLLECT KICK NET SAMPLES FOR THE REACHWIDE COMPOSITE SAMPLE**

18. Thoroughly rinse the net with stream water before proceeding to the next sampling location. It is also extremely important that all equipment, including waders, be cleaned between sites to avoid transmission of nonnative invasive species.

*Source:* Peck et al. 2006. *Environmental Monitoring and Assessment Program-Surface Waters Western Pilot Study: Field Operations Manual for Wadeable Streams.* EPA/620/R-06/003. US Environmental Protection Agency, Office of Research and Development, Washington, DC.

flowing water). Inspect and remove any organisms found on each rock and place them into the sampling container. If the kick net cannot be used at a slack water habitat because of insufficient depth of water, spend about 30 s picking up pieces of substrate from a 0.09-m² (1-ft²) area at the sampling point. Inspect and remove any organisms found on each piece of substrate and place them into the sampling container. At vegetation-choked sampling points where neither procedure can be used, sweep the net through the vegetation for 30 s, then place the contents into the sampling container.

## 6.5.2  Lake Zooplankton

The procedures recommended in this protocol for collection and preservation of lake zooplankton are based largely on the protocols designed by the EPA for the National Lakes Survey conducted in 2007 (US EPA 2007). The general procedure is described in this section. More detail is provided in Table 6.6.

Two vertical plankton tow samples are collected at the lake-sampling index site location. One sample is typically collected using a fine-mesh (typically ~50- to 80-μm) and one sample is collected using a coarse-mesh (typically ~200- to 250-μm) plankton net (Figure 6.3). We recommend an 80-μm Wisconsin net for the fine mesh and 243-μm Wisconsin net for the coarse mesh. Each net is attached to a collection bucket. Some of the larger species of zooplankton can swim fast enough to avoid being caught in the net. The coarse mesh net optimizes capture of the faster-swimming macrozooplankton because the pressure wave above the net is minimized. The fine mesh net optimizes capture of the microzooplankton. The two samples are collected and analyzed separately. Each tow is collected by pulling the sampling apparatus from a depth of about 1 m above the lake bottom to the lake surface. It is important to avoid touching lake sediments with the sampling apparatus because that can clog the net pores and compromise the integrity of the sample. The net should be raised steadily, but rather slowly (~1 ft/s), to reduce the pressure wave

**TABLE 6.6    DETAILED PROCEDURES FOR COLLECTING ZOOPLANKTON SAMPLE**

1. Fill out sample label.
2. Measure lake depth at sample location.
3. Clean and thoroughly rinse the inside surfaces of the nets and collection buckets with DIW.
4. Inspect nets and buckets for possible holes or tears.
5. Attach collection bucket to small end of each net.
6. Attach marked (every 0.5 m) lines to large end of coarse net.
7. Lower coarse net in constant upright position over side of boat until the mouth of the net is about 1.0 m above the lake bottom. If the lake depth is less than 2 m and the Secchi disk can be seen at the bottom, collect a second tow with the coarse net and combine the replicated samples (make note of this on data form).
8. Retrieve net to surface at constant rate (about 1 ft/s) without stopping.
9. At the surface, slowly move the net up and down in the water column, without submersing the net mouth, to flush zooplankton from the net into the collection bucket.
10. Further rinse contents from net into collection bucket by spraying net from outside to inside with squirt bottle containing DIW.
11. Holding collection bucket in vertical position, detach it from net.
12. Swirl the bucket without spilling the contents to filter excess water out of the bucket through the screened sides.
13. Repeat steps 6 through 12 with the fine-mesh net on the opposite side of the boat.

*Source:* Modified from US Environmental Protection Agency. 2007. *Survey of the Nation's Lakes. Field Operations Manual.* EPA 841-B-070004. US Environmental Protection Agency, Washington, DC.

that can build up at the top of the net during retrieval. Some species can detect this wave and swim out of the path of the net. Use of a wide net aperture (30 to 50 cm) can be helpful to avoid missing fast-swimming taxa such as Chaoborus, Leptodora, and Mysis.

If the lake depth at the sampling site is less than 2.0 m and the Secchi disk is visible at the bottom, a second vertical tow is made with each net (fine and coarse mesh), and the original (first) and the second samples are combined. Note that the samples collected using the fine- and coarse-mesh nets are *not* combined. If the net or attached collection bucket touches the lake sediment, field personnel should retrieve and rinse the apparatus and repeat

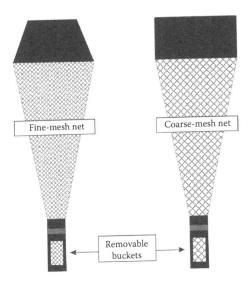

**Figure 6.3**  Wisconsin net and collection bucket diagram. Some microzooplankton nets have a reducing collar attached. (Redrawn from US Environmental Protection Agency. 2007. *Survey of the Nation's Lakes. Field Operations Manual.* EPA 841-B-07-004. US Environmental Protection Agency, Washington, DC.)

the process. Detailed description of the procedures to be followed for sample collection is provided in Table 6.6. If other mesh sizes are used instead of those that we recommend, it is important to standardize these mesh sizes such that there is consistency among sites and over time in the sampling program.

## 6.6  SAMPLE PROCESSING, PRESERVATION, AND HANDLING

### 6.6.1  Stream Benthic Macroinvertebrates

After collecting kick net samples for the reachwide samples, prepare a composite index sample from the contents of the container as described in Table 6.7. You will need to record tracking information for each composite sample on the Sample Collection Form. Check to be sure that the completed label on each jar is covered with clear tape, and that a waterproof label is placed in each jar and filled in properly. Confirm that the inside and outside labels describe the same sample. Replace the lid on each jar and seal with plastic electrical tape. It is helpful to mark the lid of each jar with the

**TABLE 6.7    PROCEDURE FOR PREPARING COMPOSITE SAMPLES FOR BENTHIC MACROINVERTEBRATES**

1. Pour off the water from the reachwide bucket through a sieve (or sieve bucket) with 500-μm mesh. Remove any large objects, such as sticks, rocks, or large plant material, from the bucket or container. Inspect these objects carefully and dislodge any clinging organisms back into the sample bucket or container before discarding.

2. Estimate the total volume of the sample in the sieve and determine the size (500-ml or 1-L) and number of jars that will be needed for the sample. Avoid using more than one jar for each of the composite samples if possible (but do not fill the jar more than a quarter full with sample).

3. Fill in a sample label with the stream ID, date of collection, and other required information. Attach the completed label to the jar and cover it with a strip of clear tape. Record the sample ID number for the composite sample on the Sample Collection Form. For each composite sample, make sure the number on the form matches the number on the label.

4. Wash the contents of the bucket or container to one side. Transfer the sample from the bucket or container into a jar, using a large-bore funnel if necessary. Use as little water from the wash bottle as possible to help transfer material. If the jar becomes too full of liquid, carefully pour off the water through the sieve. Continue to transfer sample material to the jar until it *is not more than a quarter full of solid material.* Use additional jars for the remaining sample. Carefully examine the bucket or container for any remaining organisms and use watchmaker's forceps to place them into the sample jar.

   If a second jar is needed, fill in a sample label that does not have a preprinted ID number on it. Record the ID number from the preprinted label prepared in step 4 in the *SAMPLE ID* field of the label. Attach the label to the second jar and cover it with a strip of clear tape. Record the number of jars required for the sample on the Sample Collection Form. *Make sure the number you record matches the actual number of jars used.* If possible, write *Jar N of X* on each sample label using a waterproof marker (*N* is the individual jar number, and *X* is the total number of jars for the sample).

5. Place a waterproof label with the following information inside each jar:
   Stream ID number
   Name of stream
   Date of collection
   Collector's initials

**TABLE 6.7 (*Continued*)  PROCEDURE FOR PREPARING COMPOSITE SAMPLES FOR BENTHIC MACROINVERTEBRATES**

6. Remove as much water as you can from each sample jar without removing any sample material by pouring it through the sieve. If possible, *completely fill* each jar with 95% ethanol (no headspace) so that the final concentration of ethanol is between 75% and 90%. It is important that sufficient ethanol be used or the organisms will not be properly preserved. Do not freeze samples to preserve them.

   **Note:** For backcountry work, prepared composite samples should be transported to the vehicle before adding ethanol. In that case, fill each jar with stream water and a minimal amount of ethanol to cushion the sample from the grinding action during transport of nonbiological material in the sample. Replace the water with ethanol at the vehicle as soon as possible.

7. Replace the lid on each jar. Slowly tip the jar to a horizontal position, then gently rotate the jar to mix the preservative. Do not invert or shake the jar. After mixing, seal each jar with plastic tape.

8. Store the labeled sample jars in a container with absorbent material that is suitable for use with 95% ethanol until transport or shipment to the laboratory.

site number; use a permanent marker or write on a piece of light-color tape (or a small blank address label) and attach it to the lid. Place the sample jars in a cooler or other secure container for transporting or shipping to the laboratory. The container and absorbent material placed between the jars should both be suitable for transporting ethanol. Check to see that all equipment is returned to the vehicle. Samples do not need to be kept on ice or cooled after they are preserved with ethanol.

## 6.6.2  Lake Zooplankton

After rinsing the outside of the plankton net using the squirt bottle with deionized water (DIW), transfer the sample to one (or more, if needed) sample jars. The collected zooplankton are doped by adding $CO_2$ tablets or Alka-Seltzer® to stop their movement and then preserved. Detailed procedures for sample preservation are given in Table 6.8. Zooplankton samples, once preserved, do not need to be stored on ice and can be shipped via ground transport to the laboratory. Field personnel should include one copy of the data form along with the samples (each in its own plastic bag) when the samples are shipped to the laboratory. Take one copy of the data form in the lake folder back to the office. Note that the development and use of a folder for each study lake is described in the water chemistry sampling protocol. The sample jars should be surrounded with packing material prior to shipping.

**TABLE 6.8    DETAILED PROCEDURES FOR PROCESSING ZOOPLANKTON SAMPLES**

1. Place bucket that had been attached to the coarse net into a 500-ml container filled three-fourths full with lake water to which a $CO_2$ tablet has been added (alternatively, Alka-Seltzer or club soda can be used). Wait until zooplankton have been narcotized and stop moving (about 1 min).

2. Transfer contents of bucket into a 125-ml polyethylene jar using DIW from the squirt bottle. Continue to rinse the bucket until the majority of the collected zooplankton are transferred to the jar.

3. Drain much of the excess water out of the jar by attaching a modified (to create a strainer) lid that has been prepared in advance by cutting two holes in the lid and gluing small pieces of the appropriate (large or small) mesh material to the inside of the lid to cover the holes. Carefully decant the excess water from the jar while retaining the zooplankton inside the jar.

4. Fill the jar a little more than half full with 95% ethanol.[a] If the volume of zooplankton collected fills the jar more than half full, use a second (and third if necessary) jar to preserve the additional sample volume. Record the number of jars used on the Zooplankton Sample Data Form. Label each jar identically and then add to the labels "1 of $x$,", "2 of $x$," and so on, with $x$ being the number of jars used for the sample.

5. Record the length of tows collected on the Zooplankton Sample Data Form. Verify that all required information is provided on sample labels and data form. Cover each label with clear tape.

6. Seal jar lids by wrapping with electrical tape in a clockwise direction so the lid is pulled tight as tape is stretched around it.

7. Place jar in ziplock-type plastic bag.

8. Repeat steps 1 through 7 for the second (fine-mesh) sample collected.

9. Thoroughly clean and rinse all equipment and the strainer lids before transporting them to another lake.

*Source:* Modified from US Environmental Protection Agency. 2007. *Survey of the Nation's Lakes. Field Operations Manual.* EPA 841-B-070004. US Environmental Protection Agency, Washington, DC.

[a] For backcountry sampling, add only a small amount of ethanol in the field and then fill jars to near the top with water. Discard the water and replace with ethanol on returning to the vehicle.

## 6.7 DOCUMENTATION AND TRACKING

Example labels for stream benthic macroinvertebrate samples and lake zooplankton samples are shown in Appendix D. Data collection forms for stream macroinvertebrate sampling and lake zooplankton sampling are also given in Appendix D. Labels and data collection forms should be carefully completed and double-checked before leaving the field site.

## 6.8 LABORATORY ANALYSIS OF BIOLOGICAL SAMPLES

Biological samples will need to be sent to a contract or agency laboratory where experts will enumerate and identify the individual organisms in the composite samples. Taxonomic richness results are sensitive to both counting effort (e.g., how many individuals are counted) and taxonomic resolution. Thus, it is imperative that a consistent laboratory counting protocol be used when multiple labs are involved or samples are analyzed over a period of time for trends determination. For the stream benthic IBI assessment, we recommend a 500-fixed-organism-count protocol using a gridded sorting tray (typically a 5 x 6 grid with 30 cells). Individual grid cells from the tray are selected at random and completely processed until more than 500 organisms are enumerated. The percentage of the sample processed is calculated as the number of grids processed/total number of grids, and this number is used to infer the total number and density of individuals in the composite sample from the number counted. For counts of the EPT taxa only, grid cells should be processed either until 100 individuals of the EPT taxa have been enumerated or until 500 total organisms have been enumerated.

Benthic organisms should be identified, if possible, to the genus level except for the following noninsect taxa: oligochaetes, polychaetes, and arachnids to family; nematodes and platyhelminthes to phylum. In most cases, identification of insects to family should be considered the minimal requirement. For a basic EPT taxa-richness assessment, the EPT orders should be identified to genus. For lake zooplankton, individuals should be identified to the species level if possible. Each net sample is counted independently and at least half the sample volume examined. Subsamples are examined and counted until no new species are found or until a total of 300 to 500 individuals has been counted. Either approach is acceptable; the choice should be based on the standard protocol of the laboratory.

## 6.9 QUALITY ASSURANCE

It is always advisable to replicate a portion of the samples, regardless of whether they are chemical or biological. This offers an opportunity to evaluate variability that may be introduced in the course of sampling, preserving, and analyzing the samples. Although we do not recommend that replicate zooplankton or benthic macroinvertebrate sampling should necessarily be required, we do think it is a good idea. Our recommendation is that about 5% to 10% of the sampled lakes or streams are replicated. Sample information provided on the data form for the replicate in this case will be identical to that of the original sample except for the sample ID and the time of sampling, which will differ slightly between the first and second samples at a given site. The replicate zooplankton sample should be collected at the same general location as the primary (first) sample, on the opposite side of the boat. If a stream site is to be replicated, the additional (replicate) sample is collected at each transect location to yield a pooled composite replicate, comparable to the composite normal sample. The replicated stream benthic macroinvertebrate sample at each transect location along the sample stream reach should be collected at a different stream location from the normal sample. For the replicate, move from left to center; move from center to right; move from right to left. Check the box on the data form indicating whether the sampling was replicated at this site. If the stream is not wide enough to accommodate collection of a second (replicate) sample, slide the replicate site upstream about 10 m and collect the replicate sample there. With replicate sampling, extra caution must be taken not to disturb any of the actual sampling sites by walking in them prior to sampling.

A major potential pitfall in any aquatic invertebrate study is the inherent variability and uncertainty in taxonomy among aquatic entomologists and among laboratories. This can be especially problematic in a long-term monitoring study if different laboratories or laboratory staff are involved in the identification of collected organisms over the course of the study. We recommend choosing a highly experienced taxonomic laboratory and trying to maintain consistency throughout the project. When multiple labs are involved, interlaboratory quality assurance is essential. In the EPA's national stream survey, which used eight different laboratories, 10% of the samples were randomly selected for quality control reidentification and sent to an independent taxonomist in a separate laboratory for comparison (Stribling et al. 2008). The results of the sample-based comparisons were summarized as percentage taxonomic disagreement (PTD) and percentage difference in enumeration (PDE). PDE differences among labs were minor (<3%), but PTD were on the order of 20%. Having lab taxonomists intensively interact, resolve differences, and update the data after the first round of identifications was important and improved PTD substantially in the EPA survey. We also recommend that at least 5% to 10% of samples be sent to an alternate laboratory or alternate entomologist to evaluate any differences that might arise in taxonomic identification. Such differences should be resolved, if possible, prior to finalizing the dataset.

## 6.10 INTERPRETATION.

Analysis of lake or stream water quality data can provide critical information regarding the status, or change over time, in biologically relevant water chemistry. Thus, it may be known or suspected based on measured water chemistry that in-lake or in-stream biota respond to a given concentration (or change in concentration) of ANC, pH, $Al_i$, and so on. Nevertheless, there is always some degree of uncertainty regarding the biological effects that actually occur under a given suite of water chemistry. Land managers can draw stronger inferences about biological effects if the biological resource itself is characterized or monitored. This can be important in setting target deposition loads, pursuing litigation, and evaluating damage or recovery scenarios. Therefore, there can be substantial value gained by sampling biota in addition to water chemistry.

Biological assemblage data are typically analyzed by calculating metrics from the list of the species, genera, or families identified and their abundances. For example, the number of different mayfly genera in the sample can be tallied, and this number becomes the mayfly genus richness metric. Richness metrics can be calculated for any defined taxonomic group (e.g., mayflies, rotifers, insects) as well as total sample richness. Similarly, richness can be calculated for any other autecological attributes such as functional feeding groups (shredder richness), habitat preference (swimmer richness), or tolerance to various pollutants. In addition, the same type of metrics can be calculated based on percentage of individuals in the sample (e.g., percentage mayfly individuals or percentage shredder individuals). There are also metrics to reflect overall sample diversity that are based on equations that aim to mathematically express diversity as a combination of overall sample richness (number of different taxa) and evenness (equality in the number of individuals across taxa).

A simple assessment can be made based on a single metric that is responsive to specific pollutants of interest. For example, mayflies are sensitive to pH, and mayfly taxonomic richness is therefore a good metric to use for acidic deposition studies. Total sample richness or diversity may also be used as a single overall measure of biological condition. The most robust measures of biological condition, however, require modeling or combining multiple metrics into one overall multimetric index. Application of a multimetric index or model requires some expertise in the biotic assemblage being assessed. In particular, gathering the necessary autecological information can be time consuming and require detailed knowledge of the different species that occur within the study region.

There are two major assessment approaches for quantifying whole-community biological condition: multimetric indices (e.g., IBI) or predictive modeling (e.g., the observed/expected or O/E approach). A multimetric index, such as the IBI, is developed by selecting the best 5–15 metrics that quantify condition over a suite of different aspects of biotic integrity and then summing individual metric scores into a single index of condition. Metric selection and interpretation of metric values at the sampling site are usually based on values observed at least-disturbed reference sites in similar settings. The predictive modeling approach uses reference sites to assemble lists of taxa that appear to be indicative of least-disturbed reference condition (the expected or E list). Taxa lists from a specific study site comprise the observed or O list. The proportion of the expected taxa found in the observed list (O/E ratio) is a measure of the proportion of the taxa expected to be at an undisturbed site that are actually present at the study site. An O/E ratio of 1 indicates a high-quality site (all expected taxa present). An O/E ratio of less than 0.5 means that less than half the expected reference taxa are present at the site. In practice, the E list is developed for each study site by statistical modeling (cluster analysis and discriminate function analysis) of reference site data to take into account natural differences in expected taxa distributions. This modeling approach was pioneered in Great Britain (Moss et al. 1987) and has been applied in many different locations throughout the world. These predictive models require statistical expertise to develop in new regions and a large number of reference sites as the basis for the modeling. Study site and reference site data must be collected with comparable field protocols, lab protocols, and taxonomic resolution. Because of its complexity and data requirements, we do not recommend the O/E approach as a routine tool for biological assessment on lands potentially influenced by atmospheric S and N deposition. Nevertheless, this can be a powerful tool for stream biological assessment if one is willing to develop the modeling approach and reference site database for a given region or study area.

In conducting any biological assessment, the level of taxonomic resolution (e.g., order, family, genus, or species) is an important consideration. In general, identifications to lower taxonomic levels cost more but provide more information. For stream macroinvertebrates, identification is usually taken to either the family or genus level. Some organisms can be identified to species, but in a given sample, most of the organisms can only be identified to genus because of the lack of sample keys for many taxa and the small size of early life-history stages of many of the individuals. With stream macroinvertebrates, the major laboratory effort involves picking the organisms out of the sample matrix rather than identifying them. Therefore, the laboratory cost difference between family- and genus-level analyses may not be substantial. In terms of information content, Waite et al. (2004) found that family- and genus-level stream macroinvertebrate data were similar in their ability to distinguish among the coarse impacts (e.g., most severe versus least-severe impact classes). Genus data, however, often distinguished the subtler differences in mid-Atlantic streams (e.g., mixed/moderate impacts versus high or low impacts) better than family-level data. In their analysis, acidic deposition impacts were considered a moderate impact and not a severe impact. Ordination analysis showed that both family and genus levels of analysis responded to similar suites of environmental variables. We suggest that identification to the family level can be sufficient for many bioassessment purposes. However, identifications to genus do provide more information, especially in genera-rich families like Chironomidae. Genus or finer levels of identification are important for investigating natural history, stream ecology, biodiversity, and indicator species. Decisions about the taxonomic level of identification need to be study specific and depend on available resources (cost) and study objectives.

The IBI is a multimetric index that has been used extensively in streams to characterize fish, macroinvertebrate, and periphyton condition. Note that we do not recommend application of an IBI as part of the routine process of evaluating biological response to atmospheric deposition stressors. In general, individual analyses for one or more of the EPT orders, or application of an EPT Index, is often sufficient. If, however, there is a need to more fully characterize biological conditions in a particular stream reach, and if appropriate invertebrate taxonomic and autecological expertise is available to the project team, then a stream benthos IBI is an appropriate way to proceed. Once the samples have been collected, there is not usually a dramatic difference in cost to enumerate all taxa (for implementation of an IBI) as opposed to just the insect orders included in the EPT. Nevertheless, application and interpretation of the IBI does require that more specialized taxonomic and autecological expertise be available to the project.

There are a number of different approaches to calculating IBIs, but they all follow a similar process. First, the metrics that best reflect condition are selected from the set of candidate metrics. IBIs typically have 5 to 15 different metrics. Metric values are then scored to a consistent scale (e.g., 0–10 points) and summed to calculate the one overall IBI value. A wide variety of IBIs has been developed for different stream types and regions around the world. For assessing stream benthos, we recommend as a starting point the macroinvertebrate IBI developed by the EPA for the National Wadeable Streams Assessment (WSA), in part because it was designed to be applied nationwide (Stoddard et al. 2008). The WSA IBI is formulated differently for each of nine different ecoregions in the United States. Candidate metrics were divided into six different categories, and the best-performing metric in each category was selected for inclusion in the regional IBI. The six metrics in each regional IBI are listed in Table 6.9. The six metric values were then each scored on a 0–10 scale and summed into a final IBI score (see Stoddard et al. 2008 for calculation details). There has been much less work done on developing IBIs for lake systems. We do not recommend calculating an IBI for lake zooplankton at this time as we are not aware of any existing IBIs that are ready for use.

TABLE 6.9 METRICS IN EACH CATEGORY USED IN THE EPA NATIONAL WADEABLE STREAM ASSESSMENT

| Metric Category | Individual Metrics | Aggregate Ecological Regions | | | | | | | | |
|---|---|---|---|---|---|---|---|---|---|---|
| | | NAP | SAP | CPL | UMW | TPL | NPL | SPL | WMT | XER |
| Composition | **% EPT taxa** | x | | | | | | | | |
| | % EPT individuals | | | | | | x | | x | |
| | % Noninsect taxa | | | | | x | | x | | |
| | % Noninsect individuals | | | | | | | | | x |
| | % Ephemeroptera taxa | | x | x | | | | | | |
| | % Chironomid taxa | | | | x | | | | | |
| Diversity | **Shannon diversity** | | x | x | x | x | | x | x | x |
| | % individuals in top 5 taxa | x | | | | | | | | |
| | % individuals in top 3 taxa | | | | | | x | | | |
| Feeding | **Scraper richness** | x | x | x | | x | x | x | x | x |
| | Shredder richness | | | | x | | | | | |
| Habit[a] | **% burrower taxa** | | x | | x | | x | x | | |
| | % clinger taxa | x | | x | | | | | x | x |
| | Clinger taxa richness | | | | | x | | | | |

| | | | | | | | | | |
|---|---|---|---|---|---|---|---|---|---|
| **Richness** | **EPT taxa richness** | x | x | x | x | | | x | x | x |
| | Ephemeroptera Taxa Richness | | | | x | | | | | |
| | Total Taxa Richness | | | | | x | | | | |
| **Tolerance** | **Intolerant richness** | | | | x | x | | x | | x |
| | % tolerant individuals | | x | x | | | | x | x | |
| | % PTV[b] 0–5.9 taxa | x | | | | | | | | |
| | % PTV 8–10 taxa | | | | x | x | | | | |

*Source:*  Modified from US Environmental Protection Agency. 2006. *Wadeable Streams Assessment: A Collaborative Survey of the Nation's Streams.* EPA 841-B-06-002. US Environmental Protection Agency, Office of Water, Washington, DC.

*Note:*  Metrics were selected and scored separately for each of nine aggregate ecological regions: NAP, Northern Appalachians; SAP, Southern Appalachians; CPL, Coastal Plain; UMW, Upper Midwest; TPL, Temperate Plains; NPL, Northern Plains; SPL, Southern Plains; WMT, Western Mountains; and XER, Xeric

[a]  Habit reflects the life strategy of the various taxa with respect to maintaining position in the stream (i.e., burrowing, clinging).

[b]  PTV, pollution tolerance value.

If an atmospheric deposition effects study design calls for biological characterization, for example in conjunction with a chemical characterization or monitoring effort, we recommend analysis of taxonomic richness of stream benthic macroinvertebrates or lake zooplankton for a group of streams or lakes across a gradient of acid-base chemistry. Such an analysis should be based on at least 10 water bodies, preferably more. The preferred chemical metric is usually ANC; the analysis should also be conducted for pH, $Al_i$, or $NO_3^-$ concentration. The preferred biological metric is species richness or genus richness; in some cases, family richness is the best that can be done because of taxonomic uncertainties. The taxonomic groups to be considered can include crustaceans or rotifers for lakes; mayflies, caddis flies, or stone flies for streams; or some combination of the above.

The basic data analysis for studying the effects of stressors on biological condition involves plotting biological metric scores or IBI scores versus water chemistry, as shown schematically in Figure 6.4a for mayfly genera richness. The strength of the relationship can be evaluated using the $r^2$ statistic. This analysis provides useful information on the extent to which invertebrate biological assemblages are associated with water acid-base chemistry. Trends analysis cannot be used to interpret biological change in response to improved or declining acid-base chemistry unless this basic analysis is performed and in fact yields a meaningful relationship. For water bodies that are included in long-term chemical monitoring, one should also consider subjecting at least a subset of those water bodies to biological monitoring. The biological monitoring candidates should preferably be relatively low in ANC and pH, exhibit chemistry that is not excessively variable within and among years, and exhibit reasonably rich biological assemblages. Selection of two to four waters, spread across the ANC gradient (to the extent that such a gradient occurs) between about −50 µeq/L and 50 or 100 µeq/L, would be appropriate. Resulting monitoring data should be analyzed as shown schematically in Figure 6.4b (or some variation thereof). This analysis allows determination of the extent to which chemistry and biotic richness are deteriorating or improving over time and the degree to which those trends are linked.

Variability in any of the figures used to examine relationships between stream or lake chemistry and biological community metrics can be caused by changes in environmental conditions, especially hydrological conditions. For that reason, it is always advisable to examine the influence of weather/hydrology on the observed relationships. This can easily be done by coding the points on any of these figures according to hydrological conditions (in discrete classes). This can be based on inlet or outlet stream discharge (i.e., cumulative seasonal or annual stream flow), seasonal or annual precipitation, date of snowmelt, or another variable constructed to represent the differences between wet years or seasons and dry years or seasons. This allows the analyst to determine to a first approximation the extent to which the observed relationships between chemistry and biology are influenced by hydrological differences.

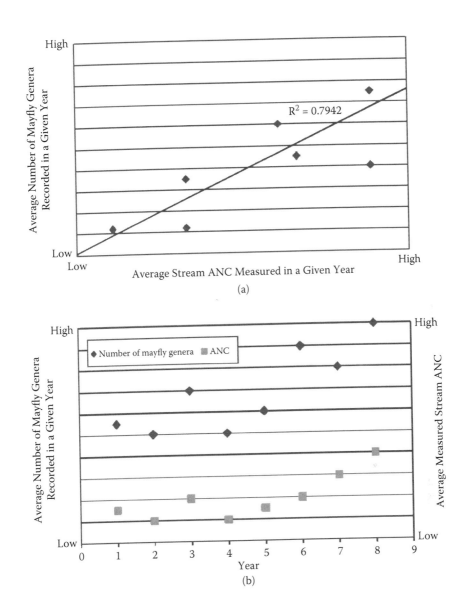

**Figure 6.4**  Schematic hypothetical examples of plots to examine mayfly richness over time in response to changes in stream chemistry plotted over a period of 8 or more years for a group of seven streams. Plot A shows richness plotted against chemistry; plot B shows both richness and chemistry plotted against time for one of the streams.

## 6.10.1   Streams

For documenting biological effects in streams in response to changes in atmospheric deposition, we recommend analyzing the quantitative relationships between invertebrate community metrics and stream ANC in multiple streams selected across an ANC gradient within a given study area. The same analyses could also be done using the variables pH, $NO_3^-$, and $Al_i$. For an initial analysis, we further suggest that for studies of response to acidic deposition, the analysis can for the sake of simplicity be limited to insects (class Insecta of the phylum Arthropoda) of the orders Ephemeroptera, Plecoptera, and Trichoptera (EPT) because of their general importance to stream ecology and their demonstrated responsiveness to changes in acid-base chemistry. We recommend examination of the number of genera (or, if that is not possible, families) present within each of these three orders, both individually and combined, in relation to differences among streams in stream chemistry. Figure 6.5 shows an example of this analysis for mayflies in streams in Shenandoah National Park (Sullivan et al. 2003). The analysis was based on both the minimum ANC and the average ANC of multiple measurements in a given stream. The same type of analysis can be conducted for a single chemistry measurement from each stream if that is what is available. The analysis shown in Figure 6.5 should be conducted for all three of the principal insect orders plus for the EPT Index. The EPT Index is calculated as either the total number of genera or the total number of families present in a given stream from the orders EPT. It represents the number of genera or families among all three orders enumerated in a single sample or the average of multiple samples. In general, we recommend basing an EPT Index on the number of genera present. If that is not possible, the calculation can be based on the number of families. The data shown in Figure 6.5 illustrate, as is often the case, that relationships between mayfly richness and stream chemistry are typically stronger than relationships for caddis flies. Stone flies alone are often not very sensitive to changes in ANC and pH.

For trends analysis of change in benthic insect diversity over time, we recommend plotting the number of genera or families (within each of the three orders individually and combined as an EPT Index) recorded for one or multiple (averaged) samplings from a given stream each year over a period of at least 8 years. This will provide an assessment of possible changes in benthic insect richness over time that can then be related to possible changes in stream ANC or some other variable. For example, the average number of genera or families of mayfly recorded during various samplings in a given year ($y$ axis) should be plotted against the average ANC ($x$ axis) determined for those same sampling occasions over the period of study. In addition, both the average number of genera or families of mayfly and the average stream ANC should be plotted

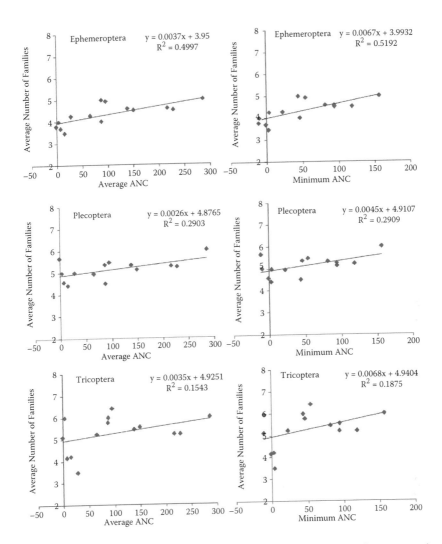

**Figure 6.5** Average number of families of aquatic insects for each of 14 streams in Shenandoah National Park versus the mean (left) or minimum (right) ANC of each stream. The stream ANC values are based on quarterly samples from 1988 to 2001. The invertebrate samples are contemporaneous. Results are presented for the orders Ephemeroptera (top), Plecoptera (center), and Tricoptera (bottom). The regression relationship and correlation are given on each diagram. (From Sullivan et al. 2003. *Assessment of Air Quality and Related Values in Shenandoah National Park*. NPS/ NERCHAL/NRTR-03/090. US Department of the Interior, National Park Service, Northeast Region, Philadelphia. http://www.nps.gov/nero/science/FINAL/shen_air_ quality/shen_airquality.html.)

over time (across the years of record) using the same time scale (cf. Figure 6.4). Such analyses allow evaluation of the extent to which changes in biota are associated with change in chemistry and the degree to which either or both are changing over time.

If the result of application of an EPT Index is not clear, that result may be attributable to a lack of invertebrate response, or it may be that the index is not sufficiently sensitive to illustrate the biological response that has occurred. In such a situation, you could consider the possibility of applying an IBI, which may be a more powerful approach.

## 6.10.2   Lakes

We recommend, for lake characterization studies focused on acidification, analyzing zooplankton data for more than 1, preferably 10 or more, lakes across an ANC gradient to determine any relationships that might exist between zooplankton richness and lake ANC. Parallel analyses can be conducted for pH and Al$_i$ in addition to ANC. These analyses should be conducted for all zooplankton groups combined (total zooplankton) and for discrete groups of zooplankton. The discrete groups should include crustaceans and rotifers at a minimum and could also include large cladocerans. An example for Adirondack lakes in New York, showing the number of zooplankton species versus ANC at the time of a zooplankton survey, is shown in Figure 6.6.

It is generally expected that variation (or scatter) in the relationships between lake chemistry and taxonomic richness may increase as the size of the study area increases. Thus, an analysis such as is shown in Figure 6.6 for a large region may yield so much variability that patterns are not clear. It may be necessary to restrict analyses such as this to a specific wilderness or to a designated subset of a region or forest, such as a certain geological type, ecoregion, or elevational band. Thus, it is advisable to examine differences in the relationships between zooplankton and lake chemistry under varying schemes for creating data subsets into groupings of lakes that are generally more similar to each other than the group of all lakes across a given region.

If there are clear relationships between zooplankton species richness and lake chemistry across a wilderness, forest, or designated subset of lakes within a wilderness, forest, or region, then it can be useful to develop a time series database for one or more presumed acid-sensitive lakes (having ANC between about −20 and +50 µeq/L). Such a database would entail contemporaneous zooplankton species richness and lake ANC measurements over a period of time of at least 8 years. Plots can then be constructed to determine, for a given lake, the relationship between ANC and zooplankton richness and changes in both of these variables over time using plots such as those depicted in Figure 6.6. We do not recommend this as a standard component of chemical long-term

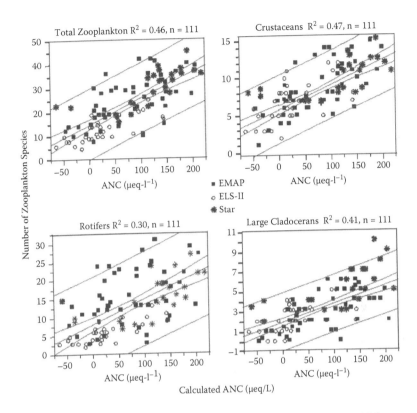

**Figure 6.6**  Total zooplankton taxonomic richness versus lakewater ANC for a combined Adirondack data set, based on 111 lake visits to 97 lakes in EPA's EMAP, Eastern Lakes Survey (ELS), and Science to Achieve Results (STAR) zooplankton surveys. (From Sullivan et al. 2006. *Assessment of the Extent to Which Intensively-Studied Lakes Are Representative of the Adirondack Mountain Region.* Final Report 06-17. New York State Energy Research and Development Authority, Albany.)

monitoring efforts. However, if it is important to document changes in biological effects in response to anticipated changes in lake chemistry, then a time series of zooplankton richness (for crustaceans, rotifers, total zooplankton, or other taxonomic grouping) may be the most straightforward and cost-effective way to do that.

Biotic assemblages in lakes vary at both temporal and spatial scales influenced by such factors as climate, vegetative cover, and disturbance. Therefore, environmental indicators exhibit variability that has a great influence on our ability to estimate biological status or trends over time. Stemberger et al. (2001) attempted to quantify the various contributions to the variance in zooplankton status as part of the EPA's EMAP sampling program in the northeastern

United States. Variance in zooplankton indicators was attributed primarily to four components of variance:

1. *Lake variance*: The lake-to-lake variability in zooplankton indicators in the study population. This depends largely on such factors as lake size, depth, fish presence/absence, pH, thermal characteristics, and productivity (Dodson et al. 2000).
2. *Year variance*: Coherent variation from year to year across all lakes, caused by, for example, an unusually warm or wet weather pattern.
3. *Lake-by-year interaction variance*: Independent year-to-year variation at each lake caused by site-specific forcing factors, such as variation in nutrient inflows or mixing regime.
4. *Index variance*: Local spatial and temporal variance caused by, for example, within-index period temporal changes, measurement error, or differences among crews or laboratories in application of the protocols (Stemberger et al. 2001).

In general, Stemberger et al. (2001) found lake variance to be the largest component for zooplankton in the northeastern United States, followed by index variance. Efforts to reduce the magnitude of these factors that contribute to zooplankton variance can maximize one's ability to detect differences or trends in the data.

## REFERENCES

Bailey, R.C., R.H. Norris, and T.B. Reynoldson. 2004. *Bioassessment of Freshwater Ecosystems Using the Reference Condition Approach.* Klewer Academic, New York.

Barbour, M.T., J. Gerritsen, B.D. Snyder, and J.B. Stribling. 1999. *Rapid Bioassessment Protocols for Use in Streams and Wadeable Rivers: Periphyton, Benthic Macroinvertebrates, and Fish.* 2nd ed. EPA/841-B-99-002. US Environmental Protection Agency, Office of Water, Assessment and Watershed Protection Division, Washington, DC.

Clarke, R.T., J.F. Wright, and M.T. Furse. 2003. RIVPACS models for predicting the expected macroinvertebrate fauna and assessing the ecological quality of rivers. *Ecol. Model.* 160:219–233.

Dodson, S.I., S.E. Arnott, and K.L. Cottingham. 2000. The relationship in lake communities between primary productivity and species richness. Ecology 81:2662–2679.

Elser, J.J., T. Andersen, J.S. Baron, A.-K. Bergström, M. Jansson, M. Kyle, K.R. Nydick, L. Steger, and D.O. Hessen. 2009. Shifts in lake N:P stoichiometry and nutrient limitation driven by atmospheric nitrogen deposition. *Science* 326:835–837.

Feldman, R.S. and E.F. Connor. 1992. The relationship between pH and community structure of invertebrates in streams of the Shenandoah National Park, Virginia, USA. *Freshw. Biol.* 27:261–276.

Gerritsen, J., W.E. Carlson, D.L. Dycus, C. Faulkner, G. Gibson, J. Harcum, and S.A. Markowitz. 1998. *Lake and Reservoir Bioassessment and Biocriteria.* EPA 841-B-98-007. US Environmental Protection Agency, Washington, DC.

Griffith, M.B., B.H. Hill, F.H. McCormick, P.R. Kaufmann, A.T. Herlihy, and A.R. Selle. 2005. Comparative application of indices of biotic integrity based on periphyton, macroinvertebrates, and fish to southern Rocky Mountain streams. *Ecol. Indicat.* 5:117–136.

Guildford, S. and R.E. Hecky. 2000. Total nitrogen, total phosphorus, and nutrient limitation in lakes and oceans: Is there a common relationship? *Limnol. Oceanogr.* 45:1213–1223.

Kauffman, J.W., L.O. Mohn, and P.E. Bugas Jr. 1999. Effects of acidification on benthic fauna in St. Marys River, Augusta County, VA. *Banisteria* 13:183–190.

Kerans, B.L. and J.R. Karr. 1994. A benthic index of biotic integrity (B-IBI) for rivers of the Tennessee Valley. *Ecol. Appl.* 4:768–785.

Klemm, D.J., K.A. Blocksom, W.T. Thoeny, F.A. Fulk, A.T. Herlihy, P.R. Kaufmann, and S.M. Cormier. 2002. Methods development and use of macroinvertebrates as indicators of ecological conditions for streams in the Mid-Atlantic Highlands region. *Environ. Monitor. Assess.* 78:169–212.

Klemm, D.J., K.A. Blocksom, W.T. Thoeny, F.A. Fulk, A.T. Herlihy, P.R. Kaufmann, and S.M. Cormier. 2003. Development and evaluation of a macroinvertebrate biotic integrity index (MBII) for regionally assessing Mid-Atlantic Highlands streams. *Environ. Manage.* 31:656–669.

Larsen, D.P. and A.T. Herlihy. 1998. The dilemma of sampling streams for macroinvertebrate richness. *J. North Am. Benthol. Soc.* 17:359–366.

Li, J., A.T. Herlihy, W. Gerth, P.R. Kaufmann, S.V. Gregory, S. Urquhart, and D.P. Larsen. 2001. Variability in stream macroinvertebrates at multiple spatial scales. *Freshw. Biol.* 46:87–97.

Melack, J.M., S.C. Cooper, T.M. Jenkins, L. Barmuta Jr., S. Hamilton, K. Kratz, J. Sickman, and C. Soiseth. 1989. *Chemical and Biological Characteristics of Emerald Lake and the Streams in Its Watershed, and the Response of the Lake and Streams to Acidic Deposition.* Final Report to the California Air Resource Board, Contract A6-184-32, Sacramento, CA. Marine Science Institute and Department of Biological Sciences, University of California.

Moss, D.M., M.T. Furse, J.F. Wright, and P.D. Armitage. 1987. The prediction of the macroinvertebrate fauna of unpolluted running-water sites in Great Britain using environmental data. *Freshw. Biol.* 17:41–52.

Patil, G.P., S.D. Gore, and A.K. Sinha. 1994. Environmental chemistry, statistical modeling, and observational economy. In C.R. Cothern and N.P. Ross (Eds.), *Environmental Statistics, Assessment, and Forecasting.* Lewis, Boca Raton, FL, pp. 57–97.

Peck, D.V., A.T. Herlihy, B.H. Hill, R.M. Hughes, P.R. Kaufmann, D.J. Klemm, J.M. Lazorchak, F.H. McCormick, S.A. Peterson, P.L. Ringold, T. Magee, and M. Cappaert. 2006. *Environmental Monitoring and Assessment Program-Surface Waters Western Pilot Study: Field Operations Manual for Wadeable Streams.* EPA/620/R-06/003. US Environmental Protection Agency, Office of Research and Development, Washington, DC.

Plafkin, J.L., M.T. Barbour, K.D. Porter, S.K. Gross, and R.M. Hughes. 1989. *Rapid Bioassessment Protocols for Use in Streams and Rivers: Benthic Macroinvertebrates and Fish.* EPA 440-89-001. US Environmental Protection Agency, Office of Water Regulations and Standards, Washington, DC.

Reynolds, L., A.T. Herlihy, P.R. Kaufmann, S.V. Gregory, and R.M. Hughes. 2003. Electrofishing effort requirements for assessing species richness and biotic integrity in western Oregon streams. *N. Am. J. Fish. Manage.* 23:450–461.

Reynoldson, T.B., D.M. Rosenburg, and V.H. Resh. 2001. Comparison of models predicting invertebrate assemblages for biomonitoring in the Fraser River catchment, British Columbia. *Can. J. Fish. Aquat. Sci.* 58:1395–1410.

Robison, E.G. 1998. Reach Scale Sampling Metrics and Longitudinal Pattern Adjustments of Small Streams. PhD thesis, Oregon State University, Corvallis.

Roth, N.E., J.H. Vølstad, G. Mercurio, and M.T. Southerland. 2002. *Biological Indicator Variability and Stream Monitoring Program Integration: A Maryland Case Study.* EPA 903/R-02/008. US Environmental Protection Agency, Office of Environmental Information, Fort Meade, MD.

Schindler, D.W., R.E. Hecky, D.L. Findlay, M.P. Stainton, B.R. Parker, M.J. Paterson, K.G. Beaty, M. Lyng, and S.E.M. Kasian. 2008. Eutrophication of lakes cannot be controlled by reducing nitrogen input: results of a 37-year whole-ecosystem experiment. *Proc. Natl. Acad. Sci. U S A* 105(32):11254–11258.

Stemberger, R.S., D.P. Larsen, and T.M. Kincaid. 2001. Sensitivity of zooplankton for regional lake monitoring. *Can. J. Fish. Aquat. Sci.* 58:2222–2232.

Stoddard, J.L., A.T. Herlihy, D.V. Peck, R.M. Hughes, T.R. Whittier, and E. Tarquinio. 2008. A process for creating multi-metric indices for large scale aquatic surveys. *J. North Am. Benthol. Soc.* 27:878–891.

Stribling, J.B., K.L. Pavlik, S.M. Holdsworth, and E.W. Leppo. 2008. Data quality, performance, and uncertainty in taxonomic identification for biological assessments. *J. North Am. Benthol. Soc.* 27:906–919.

Sullivan, T.J., B.J. Cosby, J.A. Laurence, R.L. Dennis, K. Savig, J.R. Webb, A.J. Bulger, M. Scruggs, C. Gordon, J. Ray, H. Lee, W.E. Hogsett, H. Wayne, D. Miller, and J.S. Kern. 2003. *Assessment of Air Quality and Related Values in Shenandoah National Park.* NPS/NERCHAL/NRTR-03/090. US Department of the Interior, National Park Service, Northeast Region, Philadelphia. http://www.nps.gov/nero/science/FINAL/shen_air_quality/shen_airquality.html.

Sullivan, T.J., C.T. Driscoll, B.J. Cosby, I.J. Fernandez, A.T. Herlihy, J. Zhai, R. Stemberger, K.U. Snyder, J.W. Sutherland, S.A. Nierzwicki-Bauer, C.W. Boylen, T.C. McDonnell, and N.A. Nowicki. 2006. *Assessment of the Extent to Which Intensively-Studied Lakes Are Representative of the Adirondack Mountain Region.* Final Report 06-17. New York State Energy Research and Development Authority, Albany.

US Environmental Protection Agency. 2006. *Wadeable Streams Assessment: A Collaborative Survey of the Nation's Streams.* EPA 841-B-06-002. US Environmental Protection Agency, Office of Water, Washington, DC.

US Environmental Protection Agency. 2007. *Survey of the Nation's Lakes. Field Operations Manual.* EPA 841-B-070004. US Environmental Protection Agency, Washington, DC.

Van Sickle, J. and S.G. Paulsen. 2008. Assessing the attributable risks, relative risks, and regional extents of aquatic stressors. *J. North Am. Benthol. Soc.* 27:920–931.

Waite, I.R., A.T. Herlihy, D.P. Larsen, N.S. Urquhart, and D.J. Klemm. 2004. The effects of macroinvertebrate taxonomic resolution in large landscape bioassessments: example from the Mid-Atlantic Highlands, USA. *Freshw. Biol.* 49:474–489.

# Chapter 7

## Transition Plan

### 7.1 BACKGROUND

It must be recognized that there are risks associated with changing the protocols of an ongoing sampling program. In some cases, there may be a substantial period of record established for a particular lake or stream that is based on preexisting protocols. Multiple waters within a particular study area may already have been surveyed and characterized with a particular set of protocols that may differ in important ways from newly developed protocols. A change in approach may introduce bias into future efforts to examine patterns in water chemistry across time or across space. Therefore, changes must be carefully considered, and the likely results of those changes (if any) must be evaluated prior to making a wholesale change in sampling or analysis methods. This transition plan provides a framework for considering such protocol changes, and their likely effects on the resulting data, prior to full implementation of any new protocols.

### 7.2 TRANSITION STEPS

This transition plan is divided into a sequential series of steps to be followed to ascertain the likelihood that protocol changes might introduce bias into ongoing monitoring or characterization efforts. In some cases, the preferred approach might be to continue to monitor surface waters using existing protocols or to augment earlier ongoing protocols with additional elements from the new protocols while retaining the basics of the existing protocols. In other cases, it might be best to transition to the new protocols after first evaluating the ramifications of methods changes.

**Step 1.** *Read and become familiar with the new protocols.* Each field technician should read the field-sampling protocol recommended here or those adopted by your research program. Each person who is involved with the analysis, quality assurance (QA), or interpretation of lake and stream water data should read all relevant protocols (field sampling, laboratory, QA/QC [quality control], data analysis, biology). Each person should become thoroughly familiar with the portions of the new protocols that are relevant to his or her work duties and should review existing procedures in local survey and monitoring programs to determine all significant ways in which the new procedures differ from past and ongoing practices.

**Step 2.** *Attend a field-sampling protocol training session.* Each person who is involved with the collection of water samples in the field should attend a training session to receive hands-on classroom and field training in the new field-sampling protocols. A set of PowerPoint slides is available to assist in this effort.

**Step 3.** *Determine if any existing lake- or stream-monitoring sites will be dropped from the sampling program or if any new sites will be added.* We emphasize the need for matching sampling sites (and sampling schedules) with research questions and needs. Project managers should review existing monitoring programs to determine whether, and to what extent, the sites being sampled provide information required to achieve program goals. In some cases, a lake or stream may be included in a long-term monitoring effort, but available data might indicate that the water body is not sensitive to the stressors of concern, receives substantial inputs of geological sulfur (which confound evaluation of effects of atmospheric sulfur deposition), or is impacted by some form of disturbance to such an extent that it is not possible to ascertain the influence of air pollutants. Thus, the decision could be made to drop one or more sites from the monitoring program or to add others that might better meet program needs. However, before dropping a site from a long-term monitoring effort, one must carefully weigh the value of data that have been collected from that site to date versus the benefit of replacing that site with a new site that may have little or no data associated with it but will provide information that is more useful for the program in the future.

**Step 4.** *Determine if sample collection protocols need to be changed.* The new protocols could involve changes in any or all of the elements of where, what, when, and how to collect water samples. A change in collection location (where) could affect the data even if this location is still considered to represent the same site. For example, a sample collected from a lake outlet could provide different data than a sample from the upper water column at the deepest portion of that same lake. If new field measurements or types of sample collection are added (what), the new procedures should be evaluated to ensure that they will not interfere with the prior field measurement and sample collection procedures. For example, if the existing sample collection location in a stream is downstream from the cross section where you decide to begin flow velocity measurements, the water sample should be collected before the velocity measurements are taken to ensure that the water sample is not contaminated with suspended solids caused by stream wading in the process of measuring velocity. Changing the time of year or frequency (when) that the sample is collected could change the data record. Switching from spring sampling to summer sampling could bias the data toward lower-flow conditions that might prevail during summer and that are typically less acidic than higher spring flow conditions. Switching from weekly sampling to monthly sampling would lower the sensitivity for detecting long-term trends. The procedures used to collect the samples (how) could also change the data. Water-sampling devices that integrate flow or depth can produce different results than dipping a bottle. Collection and transport via syringe may yield different values for some parameters (i.e., pH, dissolved inorganic carbon [DIC]) than collection and transport via bottle.

Influences such as those described should be evaluated in conjunction with changing methods in the middle of a monitoring program to determine if a sampling bias will be introduced. This should be done by performing the collection procedures with both the existing and the new procedures for a length of time that accounts for the full variation in sampling conditions. In many cases, this could require a year or more of duplicating procedures. Results obtained using the original protocol should be compared to results obtained using the new protocol, using a scatterplot with a 1:1 line added. If one approach yields results that are consistently either higher or lower than the other approach, the data points will plot consistently either above or below the 1:1 line. If there is no bias introduced by the change in protocol, the data points will be approximately evenly distributed above and below the 1:1 line. If it is determined that there is a bias introduced by the method change, then a decision will need to be made regarding whether to

1. Stick with the original protocol.
2. Shift to the new protocol and ignore the difference if it is judged that the difference is too small to be of consequence to the intended use of the data.
3. Develop a regression approach to "correct" the data points obtained using the original method to more closely approximate the results obtained with the new (and presumably improved) method.

This is a judgment call, and any of these options can be reasonable depending on circumstances.

Changes in the manner in which the sample is collected that improve precision without adding bias, such as additional steps to prevent possible sample contamination during sampling, should not influence the decision of whether, or when, to shift to the new protocols. An example of this type of change would be instituting the use of latex gloves during the sample collection. Any sampling-procedure change that is expected to reduce the likelihood of sample contamination should be viewed as a positive step that should be implemented as soon as is practical.

**Step 5.** *Evaluate the need for change in chemical measurements done in the laboratory or a change in the data quality objectives (DQOs) for the methods used.* If a new measurement is needed, the method should be fully evaluated to ensure that the desired results will be obtained. Also, if it is determined that a laboratory is unable to meet the DQOs needed for characterization or for a monitoring program, an alternative laboratory will need to be found. If the laboratory needs to upgrade or replace existing instrumentation or modify existing standard operating procedures (SOPs), it should provide duplicate results using samples that are representative of the relevant study sites for a minimum of 100 samples to document that the changes have not introduced a bias. Because of potential method interferences caused by the mix of chemical constituents in a water sample, a laboratory methods change might alter results for one type of surface water but not another.

**Step 6.** *Conduct side-by-side sampling or analysis to compare results obtained using initial protocols with results obtained using new protocols.* Such side-by-side comparisons should be conducted when potentially significant changes are made in either field or laboratory protocols that could affect long-term continuous records. In this situation, the data measured with previous protocols will need to be combined with the new data without any bias. Observed bias could be misinterpreted as real change over time. The time frame over which the side-by-side comparisons should be conducted should include the full range of variability in the parameters of interest. For combining long-term records for trends analysis, the minimum recommended length of time is 1 year. However, the length of time required for duplicate analyses is affected by the sampling frequency. Running duplicate sampling and analysis for 1 year might be adequate for a weekly sampling program, but not for a seasonal sampling program that collects only four times per year. Our overall recommendation is a minimum of 1 year and a minimum of 100 samples distributed across the various sampling sites. The side-by-side comparisons should be evaluated after sufficient data have been collected. The original protocol should not be dropped until the analysis of the side-by-side comparison is completed and the results indicate that the datasets based on the new and the old protocols can be combined without creating artifacts in the record.

**Step 7.** *Provide proper documentation to eliminate ambiguity in protocol applications.* Clearly label samples collected with the new protocols, and samples collected with the old protocols, in such a way that the differences are documented and unambiguous. For example, if the decision is made to replace bottle sampling with syringe sampling for the measurement of water pH, procedures must be in place to document this change in the database. For example, the documentation for that sample in the collection method field of the database would indicate "lab pH—bottle" or "lab pH—syringe." Details of each sampling method should also be documented in the field-sampling protocols.

## 7.3   DECISION OF WHETHER TO CHANGE PROTOCOLS

You should be careful about methods changes that potentially could influence chemical results while in the midst of an inventory or monitoring program. Especially if trends analyses are planned for the resulting data, it is always important to "compare apples with apples." In many cases, existing protocols, while not necessarily the preferred way of doing things, might best be left in place throughout the duration of a multiyear survey or of a long-term monitoring effort.

You must recognize that a methods change with unquantified impacts on sampling results will compromise your ability to make comparisons across space or across time. Especially if there already exists a long period of monitoring record, potential methods changes must be carefully considered and thoroughly documented with side-by-side sampling and analysis to preserve the integrity of future comparison studies. If it is determined that a change in protocols will require an adjustment of data values obtained using the original protocol, then you should consult with a statistician or scientist well versed in this kind of data adjustment.

# Appendix A: Protocols, Guidance Documents, and Methods Manuals Reviewed

Baker, J.R., A.T. Herlihy, S.S. Dixit, and R. Stemberger. 1997. Section 7. Water and sediment sampling. In J.R. Baker, D.V. Peck, and D.W. Sutton (Eds.), *Environmental Monitoring and Assessment Program Surface Waters: Field Operations Manual for Lakes.* Report No. EPA/620/R-97/001. US Environmental Protection Agency, Washington, DC.

Baker, J.R. and D.V. Peck. 1997. Section 4. Lake verification and index site location. In J.R. Baker, D.V. Peck, and D.W. Sutton (Eds.), *Environmental Monitoring and Assessment Program Surface Waters: Field Operations Manual for Lakes.* Report No. EPA/620/R-97/001. US Environmental Protection Agency, Washington, DC.

Baker, J.R., D.V. Peck, and D.W. Sutton (Eds.). 1997. *Environmental Monitoring and Assessment Program Surface Waters: Field Operations Manual for Lakes.* Report No. EPA/620/R-97/001. US Environmental Protection Agency, Washington, DC.

Berg, N. and S. Grant. 2006a. *Project LAKES (Lake AlKalinity Evaluation in the Sierra Nevada). Long-term (Mid-lake) Water Quality Monitoring and Zooplankton Data Collection Protocol.* June 2004, Revised May 2006. USDA Forest Service, Pacific Southwest Region, Vallejo, CA.

Berg, N. and S. Grant. 2006b. *Project LAKES (Lake AlKalinity Evaluation in the Sierra Nevada). Long-term (Mid-lake) Synoptic (Lake Outlet) Water Quality Data Collection Protocol.* June 2004, Revised May 2006. USDA Forest Service, Pacific Southwest Region, Vallejo, CA.

Eilers, J. 2007. *Guidelines for Monitoring Air Quality Related Values in Lakes and Streams in National Forests.* Draft report to the USDA-Forest Service Air Program, Ft. Collins, CO. MaxDepth Aquatics, Bend, OR.

Elias, J., R. Axler, and E. Ruzycki. 2008. *Water Quality Monitoring Protocol for Inland Lakes, Great Lakes Inventory and Monitoring Network.* Natural Resources Report NPS/ MWR/GLKN-2008/109. National Park Service, US Department of the Interior. http://science.nature.nps.gov/im/units/GLKN/Protocol/GLKN_LakesWQ_ final200806.pdf.

Herlihy, A.T. 1997. Section 9. Final lake activities. In J.R. Baker, D.V. Peck, and D.W. Sutton (Eds.), I. Report No. EPA/620/R-97/001. US Environmental Protection Agency, Washington, DC.

Merritt, G.D., V.C. Rogers, and D.V. Peck. 1997. Section 3. Base site activities. In J.R. Baker, D.V. Peck, and D.W. Sutton (Eds.), *Environmental Monitoring and Assessment Program Surface Waters: Field Operations Manual for Lakes.* Report No. EPA/620/R-97/001. US Environmental Protection Agency, Washington, DC.

Oakley, K.L., L.P. Thomas, and S.G. Fancy. 2003. Guidelines for long-term monitoring protocols. *Wildlife Soc. Bull.* 31(4):1000–1003.

Peck, D.V., A.T. Herlihy, B.H. Hill, R.M. Hughes, P.R. Kaufmann, D.J. Klemm, J.M. Lazorchak, F.H. McCormick, S.A. Peterson, P.L. Ringold, T. Magee, and M.R. Cappaert. 2006. *Environmental Monitoring and Assessment Program—Surface Waters. Western Pilot Study: Field Operations Manual for Wadeable Streams.* EPA/620/R-06/003. US Environmental Protection Agency, Office of Research and Development, Washington, DC.

Rantz, S.E. and others. 1982. *Measurement and Computation of Streamflow: Volume 1. Measurement of Stage and Discharge.* US Geological Survey Water-Supply Paper 2175. US Government Printing Office, Washington, DC. http://pubs.usgs.gov/wsp/wsp2175/pdf/WSP2175_vol1a.pdf.

Sharrow, D., D. Thoma, K. Wynn, and M. Beer. 2007. *Water Quality Vital Signs Monitoring Protocol for Park Units in the Northern Colorado Plateau Network.* Prepared for Northern Colorado Plateau Network Inventory and Monitoring Program. National Park Service, US Department of the Interior. http://science.nature.nps.gov/im/units/ncpn/Bib_library/water_quality/NCPN_Water_Quality_Protocol.pdf.

Sullivan, T.J. and A.T. Herlihy. 2007. *Air Quality Related Values and Development of Associated Protocols for Evaluation of the Effects of Atmospheric Deposition on Aquatic and Terrestrial Resources on Forest Service Lands.* Final report prepared for the USDA Forest Service. E&S Environmental Chemistry, Corvallis, OR.

Webb, J.R., T.J. Sullivan, and B. Jackson. 2004. *Assessment of Atmospheric Deposition Effects on National Forests. Protocols for Collection of Supplemental Stream Water and Soil Composition Data for the MAGIC Model.* Report prepared for USDA Forest Service, Asheville, NC. E&S Environmental Chemistry, Corvallis, OR.

# Appendix B: Basic Standard Operating Procedures for Stream Field-Sampling Activities

## CONTENTS

The purpose of this appendix is to provide basic standard operating procedures (SOPs) for field sampling focused on measurement of stream chemistry as influenced by atmospheric deposition. The recommended protocol featured in the main body of this book provides guidelines regarding how to implement a field-sampling program for water chemistry, with explanation of some of the reasons why certain steps and precautions are recommended. An SOP is a detailed explanation of sequential steps to be taken in carrying out the water sampling. The SOP is based on the principles outlined in the protocol. In some cases, there are multiple "correct" ways to carry out a component of the field sampling. The basic principles remain the same, and these are reflected in the protocol. Nevertheless, the exact steps may differ and yet still satisfy the aims of the protocol. Thus, these basic SOPs may be modified by a particular sampling program or field office, as appropriate to local conditions. In modifying aspects of the SOPs, the guidelines represented in the protocol should always be carefully considered.

## B.1  INTRODUCTION

This SOP provides guidelines for stream sampling. It is intended as a base SOP, suitable for adoption as a stand-alone procedure or for modification to fit local program needs. It is divided into individual sections that cover pretrip activities, sampling site documentation, stream sampling and sample handling, measurement of stream discharge, post-trip activities, and needed equipment and supplies.

## B.2  PRETRIP ACTIVITIES

### B.2.1  Background

Field teams conduct a number of activities in their office or at a base site. These include tasks that must be completed both before departure to the sampling site and after return from the site. This section describes pretrip procedures for office and base site activities that should be carried out in support of stream sampling.

Predeparture activities include development of sampling itineraries, instrument calibration if appropriate, equipment checks and repair, development of supply inventories, and sample container preparation. Procedures for these activities are described in the sections that follow. An example checklist for materials and supplies is given in Table B.1. Use this checklist to ensure that equipment and supplies are organized and available at the stream site in order to conduct the activities efficiently. Remember to take any safety equipment required by your organization (radios, cell phones, etc.).

Before leaving the base location, package the sample containers (typically two sample bottles and two 60-ml syringes for each site to be sampled, plus backup bottles and syringes in the event that one is lost or contaminated). Take plastic containers to transport filled syringes from the field to the laboratory. Fill out a set of water chemistry sample labels. Attach a completed label to each sample bottle and syringe. Make sure the syringe labels do not cover the volume gradations on the syringe. Place each sample bottle in a separate zipper-lock bag. Finally, make sure that ice or refrigerant for shipment to the lab is frozen or freezing so that it will be ready when you return from the field to the base location with the samples.

## B.2.2   Daily Itineraries

Field-sampling efforts should include a project leader who guides activities in the field and a project coordinator who remains in the office during the sampling effort. The project leader reviews each site folder to ensure that it contains the appropriate maps, contact information, copies of access permission letters (if needed), and access instructions. Additional activities can include confirming the best access routes, calling landowners or local contacts (if applicable), confirming lodging or camping plans and locations (with directions), and coordinating rendezvous locations with individuals who must meet with field teams prior to accessing a site. This information is used to develop an itinerary. The project leader should provide the project coordinator with a schedule for each day of sampling. Schedules include departure time, estimated duration of sampling activities, routes of travel, and estimated time of arrival at the sampling sites and return to the base site. Changes that might be made to the itinerary should be relayed by the project leader to the project coordinator as soon as possible. Miscommunications can result in the initiation of expensive search-and-rescue procedures and disruption of carefully planned schedules.

## TABLE B.1    CHECKLIST OF MATERIALS AND SUPPLIES FOR STREAM-SAMPLING SITE VISITS

| Standard Items | √[a] |
|---|---|
| Collection permits and entry permits, if required | |
| Site documentation forms (for new sites) | |
| Clipboard | |
| Site documentation reports (compiled as folders for existing sites) | |
| Stream-Water-Sampling Record Forms | |
| Insulated container with ice or frozen refrigerant (packed in sealed plastic bags or other containers) | |
| Small insulated container (with ice) for hike-in sites | |
| Watch for recording time | |
| Digital field camera with free memory and extra charged battery | |
| GPS unit with extra batteries | |
| Compass | |
| Field thermometer (with string attached) | |
| Preprocessed sample bottle with completed sample label attached; include a second bottle if sampling at that site is to be replicated; put each bottle in a clean plastic zip-lock bag | |
| Plastic gloves in sealed plastic bag | |
| 60-ml plastic syringes (with Luer-type tip) with completed sample labels attached; plastic container with snap-on lid to hold filled syringes | |
| Syringe valves (Mininert® with Luer-type adapter, or equivalent, available from a chromatography supply company) | |
| Water chemistry labels (if not already filled out and attached to sample containers at base site) | |
| Soft-lead pencils and write-in-rain-type pens for filling out field data forms and notebook entries | |
| Fine-tip indelible markers for filling out labels | |
| Roll or box of tape strips | |
| Field operations and methods documents | |
| First aid kit | |
| Backpack | |
| Extra zip-lock bags | |

*Continued*

**TABLE B.1 (*Continued*)    CHECKLIST OF MATERIALS AND SUPPLIES FOR STREAM-SAMPLING SITE VISITS**

| Optional Items (May Be Required for Specific Studies) | ✓[a] |
|---|---|
| 60-ml glass bottles with septum caps and with completed sample labels attached | |
| Calibrated multiparameter sonde, data logger and cable, with extra batteries | |
| Calibration standards, quality control check samples, deionized water, rinse bottles, waste tray and container, calibration cup, and sensor guard for sonde (multiple sensors combined in a unit that is lowered into the water) | |
| Sonde calibration and postcalibration record forms | |
| Measurement tape | |
| Waders or high-top waterproof boots for wading | |
| Clear packaging tape to cover labels | |
| Dissolved oxygen (DO)/temperature meter with probe | |
| DO repair kit containing additional membranes and probe-filling solution | |
| Conductivity meter with probe | |

[a] Check off each item as it is added.

## B.2.3   Instrument Checks and Calibration

If appropriate, each field team should test or calibrate field instruments prior to departure for the sampling site. Such testing may be appropriate for dissolved oxygen meters, global positioning systems (GPSs), and perhaps other instrumentation. Batteries should be checked prior to departure for field sites. Extra batteries should be carried.

Field personnel should check the inventory of supplies and equipment prior to departure using project-specific site visit checklists. Meters, probes, and sampling gear should be packed for transport to the field in such a way that it minimizes physical shock and vibration during transport. Rafts or float tubes should be packed for transport to minimize the potential for puncture by any sharp object.

## B.3   SITE DOCUMENTATION

### B.3.1   Background

This section describes SOPs for establishing and documenting sampling sites on small, well-mixed streams or lake outlets. This procedure applies to new sites for which approximate locations have been designated based on program objectives and sampling design. It also applies to previously established sites for which current or updated site documentation is needed.

## B.3.2   Objective

The objective of this procedure is to establish and document new sampling sites and to update documentation for established sites, providing

A. Site descriptions and notes
B. Travel and access descriptions and notes
C. Site coordinates obtained in the field using a GPS unit
D. Site and access-related photos
E. Placement or confirmation of numbered site tags (where applicable)

For established sites, existing site documentation will be evaluated for clarity and improved as needed based on conditions observed in the field.

## B.3.3   Material Needed for Use in Field for Site Documentation

A. Available site documentation records for previously established sites:
   1. Site location maps, topographic maps, and road maps
   2. Site descriptions and access notes
   3. Site tag numbers and tag tree descriptions (where applicable)
   4. Site coordinates
   5. Site photos
B. Preliminary site documentation for new sites:
   1. Site location maps, topographic maps, and road maps, indicating approximate site locations
   2. General site descriptions and access notes
C. General material for site documentation:
   1. Regional-scale topographic and road maps
   2. Stream-Sampling-Site Documentation Forms on waterproof paper
   3. Clipboard or field notebook and pens for use with waterproof paper
   4. GPS unit with replacement batteries
   5. Digital camera with charged battery and charged replacement battery
   6. Site tags, aluminum nails, and hammer (if applicable)
   7. Measuring tape
   8. Blaze orange material for flagging tag trees in photos (if applicable)
   9. Gate keys (if needed)
   10. Cell phone with numbers of project staff and management agency offices

## B.3.4   Sequence of Site Documentation Activities

A. Initiate the Stream-Sampling-Site Documentation Form. Enter the station ID, station name, date established or revised, forest (or other unit), and the name, affiliation, e-mail address, and phone number of the person responsible for site documentation.

B. Select or locate using GPS the specific sampling site (applies to new sites).
  1. The approximate or preliminary location of new site locations will be indicated on topographic maps. The sample collection team must still determine the exact point on the stream to be sampled.
  2. Avoid establishing sites where streams may not be well mixed, such as locations in close proximity to inflowing tributaries or braided channels. Also, avoid locations that may be influenced by run-off from disturbed areas, roads, trails, drainage ditches, or other sources of inflow. Select sites that are upstream rather than downstream of potentially altered inflow. As a general rule, select sites that are at least 25 m above or below confluence points or inflow.
  3. The best point to sample will be where the water is flowing fast or falling, where there are no eddies, and where the depth is at least 8 in. (20 cm). Ideally, the sampling point is one that can be reached while kneeling on the stream bank or on stable rocks downstream from the sampling point. If possible, avoid standing in the water to reach the sample point.

C. Obtain new coordinates at the site using the GPS.
  1. The unit position format should be set to decimal degrees (hddd. ddddd). The datum should be set to NAD83. Distance and elevation should be set to meters. The wide area augmentation system (WAAS) should be enabled.
  2. When "Mark Waypoint" is selected, the default GPS site ID (a number) should be changed to the actual site ID.
  3. Before saving the coordinates, note the estimated accuracy of the measurement and enter it on the Sampling-Site Documentation Form.
  4. Save the coordinates on the GPS and record both the coordinates and the elevation on the Sampling-Site Documentation Form. Do not rely solely on the GPS to store the coordinates.
  5. Confirm that the waypoint has been saved in the GPS unit.

D. Enter the approximate stream depth and width on the Sampling-Site Documentation Form. Enter the approximate average values for stream depth and width observed in the sampling-site area (about 5 m upstream and downstream of the sampling site) on the Stream-Sampling-Site Documentation From.

E. Enter station description information on the Sampling-Site Documentation Form.
   1. For existing sites, enter information to improve and update existing site description information.
   2. Generally describe the site, referring to proximity to landmarks (trails, bridges, tributaries, trees, landscape features, or other relatively permanent features). Provide any additional information, including detailed stream bank and stream structure descriptions that will help future sample collectors identify the site. Add any information here that might be relevant to water and stream quality, such as cleared land, roads, construction, logging, development, any earth disturbance, and so on observed above the site in the watershed or in the stream.
   3. As a general convention, the right and left sides of a stream are determined based on looking downstream. When making observations on the form, always indicate whether the observation is made looking downstream or upstream.

F. Enter travel and access directions on the Sampling-Site Documentation Form.
   1. For existing sites, improve and update existing travel and access information.
   2. Travel and access notes should be sufficient to guide future sample collectors to the site without reliance on GPS units. Not all future sample collectors will necessarily have GPS units.
   3. Access notes should refer to trails, roads, and permanent landmarks, providing distances and, where helpful, compass-based directions. Backtrack if necessary to determine distances. Linear distances and directions from the established site waypoint can be determined using the GPS unit.
   4. If a parking location is not immediately adjacent to the sampling site, use the GPS unit to obtain the coordinates for the parking location and record in the travel and access information entry area of the Sampling-Site Documentation Form.
   5. For complicated or long walk-ins, use a GPS unit to record and save a track. But again, do not rely on future sample collectors having access to a GPS unit.
   6. Sketch the route on the back of the Sampling-Site Documentation Form for scanning and saving as a JPEG image if that would be helpful.

G. Obtain site and access photos.
   1. For site documentation, if the camera allows, set the camera's picture size at 3 m. This will create picture files that are about

550–650 kb. Larger, higher-resolution files are not needed for site documentation work. Switch to higher resolution if you are taking pictures for other purposes.

2.  Photos should be obtained providing downstream-looking and upstream-looking views of the sampling site; views of the tag tree (if applicable, with blaze orange material attached); and views of other distinguishing features in relation to the site (bridges, roads, notable rocks, trees, landforms, signage, etc.) Photos should also be obtained to show important aspects of site access (parking area, forks in the trail, etc.).

3.  All photos should be listed on the form, including the file name, date, and description. Enter this information at the time that the pictures are taken. Do not rely on memory for later entry of photo descriptions. The entered description should serve as the photo caption for site documentation reporting.

## B.4   STREAM SAMPLING

### B.4.1   Background

Water chemistry data are used to characterize acid-base status and trophic condition and to classify streams based on their water chemistry. Samples for analysis of most parameters are collected into plastic bottles. Syringe samples or samples collected into glass bottles with septum caps are preferred for collection of sample aliquots for laboratory analysis of pH and dissolved inorganic carbon (DIC) where practical. Syringes and septum caps are used to protect samples from exposure to the atmosphere because the measured values for these parameters can change if the stream water sample equilibrates with atmospheric $CO_2$ subsequent to collection.

Stream samples are obtained at a single sampling location below the water surface in the portion of the stream cross section that appears visually to represent the greatest amount of flow or alternatively at midchannel in an area of flowing water. Spatial variability across the channel of a single stream is expected to be minimal in relatively small wadeable streams as compared to the variability expected among sites, so a composite water chemistry sample is not typically required.

At each stream, optional *in situ* and streamside measurements are made using field meters and recorded on standard data forms. Stream water is collected in one or more bottles and two 60-ml syringes or glass bottles with septum caps that are stored on ice in darkness and shipped or driven to the analytical laboratory as quickly as possible after collection. Overnight express mail to the laboratory is required for these samples because the syringe or glass bottle samples need to be analyzed, and some or all of the bottled sample needs to be stabilized (by filtration or acidification) within a short period of time (typically 72 h) after collection. Check with the analytical laboratory in advance of sampling regarding applicable holding times for the parameters to be measured.

This SOP describes procedures for routine sampling and data collection at water quality monitoring sites on streams. Water samples are collected for lab analysis with optional *in situ* (on-site) measurement of selected water quality parameters (water temperature, specific conductance, pH, dissolved oxygen, and turbidity) using a multiparameter instrument (sonde).

This section describes procedures to be followed for data collection at established water quality monitoring sites. The sites may be part of a synoptic sampling or be fixed long-term sample sites for which water quality data and water samples are collected on a scheduled periodic basis.

## B.4.2  Documentation of Data and Sample Collection

The Stream-Water-Sampling Record Form is used to document sample collection and record all field data. The form is used to record the following information:

A. The organization, station ID, and station name.
B. The date and arrival time for the site visit and specific times of measurements obtained.
C. The name, contact information, and affiliation of the individual who is the collector of record and responsible for protocol adherence during the site visit.
D. Suggested revisions or amendments to site documentation and travel directions.
E. A listing of site-related photographs taken, including file name, date, and descriptions.
F. Qualitative descriptions of weather, stream discharge level and appearance, and other factors that might influence water quality during the site visit.
G. Air temperature.
H. Results for all water quality data collected *in situ*, including
   1. Numerical results, units, and measurement time.
   2. Instruments used.
I. Identification of calibration and postcalibration sensor check records.
J. Results for all discharge data collected, including
   1. Location of discharge measurement site relative to the sampling and data collection site.

2. Numerical results, units, and specific time of measurement.
3. Methods identification.
4. Identification of discharge record files.
K. A listing of all samples collected, including
   1. Collection time.
   2. Types of samples collected and number of replicates.
   3. Method of delivery to analytical lab.

## B.4.3   Sequence of Activities for Data Collection

Collectors are advised to avoid entering or disturbing the stream or stream bank at, or upstream of, the collection site prior to sample collection and completion of water quality data recording. The typical sequence of activities on arrival at the sampling site is as follows:

A. Confirm the site location based on information in the Site Documentation Form, including coordinates, photos, and access notes.
B. Initiate completion of the Stream-Water-Sampling Record Form.
C. Complete Site Information and General Observations sections of the Stream-Water-Sampling Record Form.
D. Enter information needed to improve or correct the site description and travel directions provided on the Site Documentation Form.
E. Obtain any photographs needed to improve site documentation and enter file names, dates, and descriptions.
F. Note any factors (other than weather and discharge conditions) that might affect water quality (e.g., bank or upstream disturbance, debris in water).
G. Collect water samples and complete the Water Sample section of the Stream-Water Sampling Record Form. Enter any *in situ* data into that section of the Stream-Water-Sampling Record Form.
H. Complete the Chain-of-Custody Form.
I. Check to make sure that all of the information recorded on the sample labels, Chain-of-Custody Form, and Stream-Water-Sampling Record Form match.
J. Obtain discharge measurements or stage height data, if required. Indicate method, time of measurement, result, name of the record file, and location of measurement relative to the data and sample collection site in the Stage and Discharge Data section of the Stream-Water-Sampling Record Form. Note that discharge gauging may be conducted at the same time as other site visit activities if the discharge measurement site is downstream of the water quality and data collection site. Also, note that measurements of discharge or stage height are considered optional.

If desired, *in situ* (on-site) measurement of one or more parameters can be made using a multiparameter water quality sonde (handheld instrument with a probe [containing multiple sensors] that is lowered into the water and that measures various physical parameters). Such measurements might include temperature, pH, dissolved oxygen, specific conductance, or turbidity. The procedures for such *in situ* data collection will vary with the specific field instrument but in general require the following steps:

- Initiate water quality sonde field calibration and calibration checks. Record results on a water quality instrument calibration and post-calibration record form. Confirm that sensor check criteria are met. If criteria are not met, recalibrate, perform sensor maintenance, or replace sonde or sensors as needed to meet the criteria.
- Deploy the water quality sonde for the period required to obtain stabilization. Enter the results and time of measurement in the *in situ* Water Quality Data section of the Stream-Water-Sampling Record Form.

## B.4.4  Sample Collection

In the field, make sure that the labels all have the same sample ID number (bar code), and that the labels on the bottles (and syringes, if used) are securely attached. Carefully avoid disturbance of water upstream of the sampling point prior to sample collection. This means not walking in the upstream water or on upstream rocks.

Collect a water chemistry sample as described in Table B.2 from the middle of the stream channel at the sampling site unless no water is present at that location. Throughout the collection process, it is important to take precautions to avoid contaminating the sample. Wear gloves provided in the sample bag. Rinse all sample containers three times with stream water before filling them with the sample. Many streams have low ionic strength and can be contaminated easily by perspiration from hands, sneezing, smoking, insect repellent, sunscreen, or chemicals (e.g., formalin or ethanol) used when collecting other types of samples. Make sure that none of the water sample contacts your hands before going into the sample bottle or syringe. The chemical analyses conducted using the syringe samples can be affected by equilibration with atmospheric carbon dioxide; thus, it is essential that no outside air contact the syringe samples during or after collection.

Record the information from the sample label on the Stream-Water-Sampling Record Form. Note any problems related to possible contamination in the Comments section of the form.

**TABLE B.2    OVERVIEW OF STREAM SAMPLE COLLECTION PROCEDURES FOR WATER CHEMISTRY**

### Collection into Bottle

1. Select sample location in a flowing portion of the channel near the middle of the stream.

2. Put on gloves provided in the sample bag.

3. Rinse sample bottle and lid three times with stream water.

4. Fill the sample bottle completely, holding the bottle in a tilted position approximately at the midpoint between the water surface and the streambed, being careful not to disturb any sediment prior to or while collecting the sample.

5. If a septum cap is to be used, place the cap on the bottle under water.

6. Put the sample bottle into a clean plastic zipper-lock bag.

7. Place the sample bottle in a cooler (on ice or stream water) and shut the lid. This may be a soft cooler for packing out of the field to the vehicle or a hard cooler in the vehicle. If a cooler is not available, place the bottle in an opaque garbage bag and immerse it in the stream.

### Collection into Syringe

8. Rinse the syringe three times with water from the sampling location.

9. Slowly fill the syringe with sample, avoiding generation of air bubbles, until it is two-thirds to three-fourths full. This will help to ensure that the plunger remains inserted far enough into the filled syringe so that it will not be likely to become dislodged during transport.

10. Expel any air.

11. Repeat procedure using a second syringe.

12. Place the filled syringes into a plastic container for transport.

### Collection from Very Shallow Stream

If the stream is too shallow to collect a sample using standard procedures, the following approach can be used, using a new clean syringe at each site:

13. Rinse the syringe three times with stream water, downstream of sample site as usual.

14. Use the syringe to put stream water in the sample bottle and rinse the sample bottle three times.

15. Finally, use the syringe to fill the bottle to the brim with stream water at the sample site. Cap the bottle and proceed as normal.

## B.4.4.1   Sample Collection Procedure

The sample should be collected on a step-by-step basis as follows:

a. Remove the gloves from the plastic bag and put them on.
b. Remove the sample bottle from the plastic bag. Do not put the bag on the ground.
c. Check to ensure that the correct labels are affixed to each sample bottle and syringe.
d. Rinse the sample bottle in the stream at a location at least 2–3 feet downstream from the sample collection point. The bottle and cap should be rinsed three times. For each rinse, fill the bottle and then pour the rinse water over the inside of the cap, held bottom-side up in the other hand. Pour the rinse water downstream of the rinsing and sampling points and avoid stirring up streambed debris during the process.
e. After the rinsing is completed, move to the sampling point and collect the sample by submersing the tilted bottle or syringe to a depth midway between the sediment and the water surface. Fill the bottle as completely as possible. While collecting the sample, avoid stirring up streambed debris that might be collected with the sample. Try to avoid generating large bubbles in the bottle while it is being filled. Also, avoid collecting water that has come in contact with the gloves or the outside of the bottle. This can often be best achieved by sampling rapidly flowing or falling water. If it is deemed that debris may have entered the sample bottle, discard the contents (at a downstream location), rerinse the bottle (or use a clean backup bottle), and collect a new sample.
f. Immediately after collecting the sample, place the lid on the bottle (tightly) and return the bottle to its plastic bag. If a septum cap is being used, cap the bottle under the surface of the water to avoid any contact of the sample with air. Seal the bag.
g. If a sample is to be collected into a syringe, submerge a 60-ml syringe halfway into the stream and withdraw a 15- to 20-ml aliquot. Pull the plunger to its maximum extension and shake the syringe so the water contacts all surfaces. Point the syringe downstream and discard the water by depressing the plunger. Repeat this rinsing procedure two more times.
h. Submerge the syringe into the stream again and slowly fill the syringe with a fresh sample. Try not to get any air bubbles in the syringe. If more than one or two tiny bubbles are present, discard the sample and draw another one.

i. Invert the syringe (tip pointing up) and cap it with a syringe valve. Tap the syringe lightly to detach any trapped air bubbles. With the valve open, expel the air bubbles and a small volume of water, leaving the syringe between two-thirds and three-fourths full. Note that the syringe is transported only partially full to avoid dislodging the plunger during transport. Close the syringe valve. If any air bubbles were drawn into the syringe during this process, discard the sample and fill the syringe again (step h).

j. Repeat steps g through j with a second (backup) syringe. Place the syringes together in a separate plastic bag and place in a plastic container, which is then placed into the cooler (or stream water if that method of cooling is used while still in the field).

k. Complete the Stream-Water-Sampling Record Form while at the sample site.

l. Inspect all equipment and clean off any plant and animal material before moving to the next sample location. This effort ensures that introductions of nuisance species do not occur between streams. Inspect, clean, and handpick plant and animal remains from any footwear or equipment that may have contacted stream water.

## B.4.4.2   Sample Handling

a. Place the bagged sample on double-bagged ice or refrigerant immediately after collection or at least within 15 min of collection. *Note*: Do not put ice in the plastic bag that contains the sample bottle or in the plastic container that contains the syringes. Ice or refrigerant should be double bagged in plastic bags to avoid possible leakage and contamination of the samples. Samples can be held in a soft-sided cooler until returning to the vehicle.

b. The large sample cooler can be left in the collection team's vehicle. The sample can be transferred to the larger cooler on return to the vehicle.

c. For sites that are not close to road access, the collection team should make arrangements to keep the sample on ice after collection and during the return hike. One approach would be to use a small soft-pack cooler or other container that will fit in a backpack. Ice, snow, or refrigerant could be placed in a plastic bag in the cooler or container (double bag to avoid leakage and contamination of samples). Samples are transferred to the larger cooler at the vehicle.

d. The samples should be kept in the dark and on ice until delivery to the lab. The ice may need to be replenished during sample transit. Avoid letting the sample bottle float in melted ice water. Do not place the sample bottle in a refrigerator or cooler with food or in any container that is not clean. Ship the samples as soon as possible, preferably within 24 h of sampling.

e. Note that we do not recommend filtration in the field. If, however, a program filters the samples in the field for chlorophyll *a* measurement, it is important to record on the water-sampling record form the volume of water filtered. Record this information in the Notes section of the form. The filter is then sent to the analytical laboratory for determination of chlorophyll *a* content on the filter.

### B.4.4.3  Postsampling Actions

a. Record the sample ID number (bar code) on the Stream-Water-Sampling Record Form along with the pertinent stream information (stream name, site ID, date, etc.). Note anything that could influence sample chemistry (heavy rain, potential contaminants) in the Comments section. If the sample was collected at the targeted site, record an *X* in the STATION COLLECTED field. If you had to move to another part of the reach to collect the sample, place the letter of the nearest transect in the STATION COLLECTED field. Record more detailed reasons or information in the Comments section. Make sure that the record form is completely filled in.

b. Complete the Chain-of-Custody Form.

c. Check to make sure that all of the information matches on the sample labels, Chain-of-Custody Form, and Stream-Water-Sampling Record Form.

d. Transport the samples back to the vehicle in a soft cooler on ice or snow.

e. After carrying the samples to the vehicles, place the bottles and syringes in a cooler and surround with 1-gal resealable plastic bags filled with ice. Double bag to avoid getting cooling water into sample bags.

## B.4.5  Field Measurements

Determine stream temperature with a field thermometer (one that does not use mercury). Determine specific conductance and dissolved oxygen concentration using field meters (optional). Follow instructions provided in Table B.3. Record the measured values on the Stream-Water-Sampling Record Form.

Table B.4 describes the equipment-cleaning procedures. Inspect all equipment and clean off any plant and animal material. This effort helps to prevent introductions of nuisance species between streams.

**TABLE B.3    PROCEDURES FOR STREAMSIDE AND *IN SITU* CHEMISTRY MEASUREMENTS**

### Specific Conductance

1. Check the batteries and electronic functions (e.g., zero, red line) of the conductivity meter as instructed by the operating manual.

2. If you have not tested the meter at a base location recently, insert the probe into the *RINSE* container of the quality control check sample (QCCS) and swirl for 3 to 5 s. Remove the probe, shake it off gently, transfer it to the *TEST* container of QCCS, and let it stabilize for 20 s. If the measured conductivity is not within 10% of the theoretical value, repeat the measurement process. If the value is still unacceptable, do not use the meter until it can be inspected, diagnosed, and repaired.

3. Submerge the probe in an area of flowing water near the middle of the channel at the same location where the water chemistry sample was collected. Record the measured conductivity and any pertinent comments about the measurement on the Field Measurement form.

### Dissolved Oxygen and Temperature

1. Inspect the probe for outward signs of fouling and for an intact membrane. Do not touch the electrodes inside the probe with any object. Always keep the probe moist by keeping it inside its calibration chamber.

2. Check the batteries and electronic functions of the meter as described in the operating manual.

3. Calibrate the oxygen probe in water-saturated air as described in the operating manual. Allow at least 15 min for the probe to equilibrate before attempting to calibrate. Try to perform the calibration as close to stream temperature as possible (not air temperature) by using stream water to fill the calibration chamber prior to equilibration.

4. After the calibration, submerge the probe in midstream at middepth at the same location where the water chemistry sample was collected. Face the membrane of the probe upstream and allow the probe to equilibrate. Record the measured DO and stream temperature on the Field Measurement form. Record the time the DO and temperature measurement was made in 24-h units (e.g., 1423) on the field form. If the DO meter is not functioning, measure the stream temperature with a field thermometer and record the reading on the Field Measurement form along with any pertinent comments.

**Note:** Older model DO probes require a continuous movement of water (0.3 to 0.5 m/s) across the probe to provide accurate measurements. If the velocity of the stream is appreciably less than that, agitate the probe in the water as you are taking the measurement.

**TABLE B.3 (*Contiued*)  PROCEDURES FOR STREAMSIDE AND *IN SITU* CHEMISTRY MEASUREMENTS**

**Temperature Only (If No Field Meters Are Used)**

1. Place a field thermometer (±1°C accuracy) beneath the surface of the stream at the approximate depth of sample collection in an area of flowing water at or near where the water chemistry samples were collected.

2. Record the stream temperature (estimated to the nearest 0.1°C) on the Field Measurement form. Record the time the temperature measurement was made in 24-h units (e.g., 1423) on the field form, along with any pertinent comments (e.g., measurement taken in sun or shade).

**TABLE B.4  POSTSAMPLING EQUIPMENT CARE**

1. Clean any equipment that may have contacted surface water for biological contaminants. **If you are moving between sites on the same day, do this before moving to the next site.**

2. Clean and dry other equipment prior to storage. Rinse coolers with water to clean off any dirt or debris on the outside and inside.

3. Inventory equipment and supply needs and relay orders to the project coordinator.

4. Remove dissolved oxygen meters, other instrumentation, and GPS from carrying cases and set up for predeparture checks and calibration. Examine oxygen membranes of DO meters for cracks, wrinkles, or bubbles. Replace if necessary.

5. Recharge batteries overnight if possible. Replace other batteries as necessary.

6. Recheck field forms from the day's sampling activities. Make corrections and completions where possible and initial each form after review.

7. Replenish fuel.

*Source:*  Modified from Baker, J.R., D.V. Peck, and D.W. Sutton (Eds.). 1997. *Environmental Monitoring and Assessment Program Surface Waters: Field Operations Manual for Lakes.* Report No. EPA/620/R-97/001. US Environmental Protection Agency, Washington, DC.

## B.4.6  Post-trip Activities

On return to a lodging or office location after sampling, the team should review all labels and completed data forms for accuracy, completeness, and legibility. A final inspection should be made of all samples. If information is missing from the forms or labels, the project leader should attempt, if possible, to fill in the information accurately. The project leader should initial all data forms after review. If samples are missing or not properly labeled, it may be necessary to reschedule the site for another complete sampling. Other postsampling activities include inspection and cleaning of sampling equipment, inventory and sample preparation, sample shipment, and communications.

### B.4.6.1  Equipment Cleanup and Check

Inspect, clean, and handpick plant and animal remains from any vehicle, footwear, or equipment that may have contacted stream water. Also, try to avoid transfer of plant seeds from location to location.

### B.4.6.2  Shipment of Samples and Forms

Samples and forms should be shipped or transported to the analytical laboratory in as short a time as is reasonably possible after completion of data and sample collection. Call or e-mail the lab to alert it that samples are in transit and when to expect delivery. Samples should be maintained in insulated containers with refrigerant after collection and during transport. The Chain-of-Custody Record Form should be kept with the samples until they are logged in at the analytical laboratory. If samples are to be shipped to the laboratory, an overnight shipping service should be used, and shipping should be avoided when samples would be delayed by transit over a weekend or holiday period.

Samples should be shipped in coolers packed with ice. Line each shipping cooler with a large 30-gal plastic bag. Inside, package the ice separately within numerous (as many as feasible) self-sealing plastic bags and ensure that the ice is fresh before shipment. Use block ice when available or "blue ice." Block ice should be sealed in two 30-gal plastic bags. White or clear bags will allow for labeling with a dark indelible marker. Label all bags of ice as "ICE" with an indelible marker to prevent misidentification by couriers of any leakage of water as a possible hazardous material spill.

Line the shipping cooler with a 30-gal plastic bag to prepare the sample bottles and syringes for shipping. Place another garbage bag in the cooler and place the samples in the second bag. Put filled syringes or glass bottles in sturdy containers to prevent damage during transport. Ensure that all label entries are complete and close the bag of samples. Place bags of ice around it. Then, close the cooler liner (outer garbage bag). Ship water samples on the day of collection whenever possible. If that is not possible, they should be shipped the next day.

### B.4.6.3  Processing Site Documentation Data and Information

A database with reliable backup should be established for storage of site records and files, map images, and photos. Processing of site documentation data and information include the following steps:

A. Retrieve site coordinates (and any tracks) from the GPS unit using the GPS software. Delete any extra coordinate sets (waypoints) and save the file as a *.gdb file.
B. Retrieve photos from the camera.

C. Enter or revise the site record in the site documentation database.
   1. Enter site coordinates obtained in the field.
   2. Enter or revise the site description and travel and access directions.
   3. Enter or revise the tag and tag tree information as needed. If no changes were made, note that the tag placement was confirmed on the particular date. Note that tree tags may not be applied in a wilderness setting.
   4. Add new photos as JPEG images with captions to the site record.
D. Create site maps providing both detailed and broader information for access and orientation. Annotate maps and pictures with text and arrows when it would be helpful. Note that the accuracy of maps varies, and the coordinate-based point on the maps, as well as other information, may be misleading. Add clarifying notes. Save these maps as JPEG images in the site record. Add captions as appropriate.

## B.5   STREAM DISCHARGE

### B.5.1   Background

Stream discharge is equal to the product of the mean current velocity times the vertical cross-sectional area of flowing water. It reflects the volume of water per unit time that passes a particular location (line drawn at right angle to the stream channel) on the stream. Discharge measurements can be helpful for assessing trends in stream water chemistry that are sensitive to stream flow. Stream discharge information is also useful in interpreting the representativeness of water chemistry data and some physical habitat information. Water chemistry measured under unusually high or low flow is not expected to represent well the chemistry under average flow.

The location selected for measuring stream discharge should be as close as is reasonable to the location where chemical samples are collected. Variability in stream discharge within the reach of interest is expected to be small compared to variability in stream discharge among streams, so multiple determinations at a site are not required.

No single method for measuring discharge is applicable to all types of stream channels. The preferred procedure for obtaining discharge data for small streams is based on "velocity-area" methods (e.g., Rantz et al. 1982, Linsley et al. 1982). For streams that are too small or too shallow to use the equipment required for the velocity-area procedure, an alternative procedure is presented. It is based on timing the filling of a volume of water in a calibrated bucket.

## B.5.2   Velocity-Area Procedure

Because velocity and depth typically vary greatly across a stream, accuracy in field measurements is achieved by measuring the mean velocity and flow cross-sectional area of many increments across a channel (Figure B.1). Each increment gives a subtotal of the stream discharge, and the whole is calculated as the sum of these parts. Discharge measurements are made ***at only one*** *carefully chosen channel cross section within the sample reach.* It is important to choose a channel cross section that is as much like a canal as possible. A glide area with a U-shaped channel cross section that is free of obstructions provides the best conditions for measuring discharge by the velocity-area method. You may remove rocks and other obstructions to improve the cross section before any measurements are made. However, because removing obstacles from one part of a cross section affects adjacent water velocities, you must not change the cross section once you commence collecting the set of velocity and depth measurements.

The procedure for obtaining depth and velocity measurements is outlined in Table B.5 (based on Rantz et al. 1982). Record the data from each measurement in the Stream Discharge section of the Stream-Water-Sampling Record Form, giving for each measurement increment the distance from the left bank (facing downstream), water depth, measured velocity, and any required flags or notes.

## B.5.3   Timed-Filling Procedure

In channels too small for the velocity-area method, discharge can sometimes be determined directly by measuring the time it takes to fill a container of

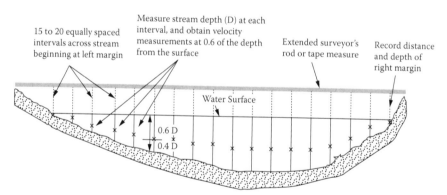

**Figure B.1**   Layout of a channel cross section for obtaining discharge data by the velocity-area procedure. (From Rantz, S.E. and others. 1982. Measurement and Computation of Streamflow: Volume 1. Measurement of Stage and Discharge. US Geological Survey Water-Supply Paper 2175. US Government Printing Office, Washington, DC. http://pubs.usgs.gov/wsp/wsp2175/pdf/WSP2175_vol1a.pdf.)

## TABLE B.5    VELOCITY-AREA PROCEDURE FOR DETERMINING STREAM DISCHARGE

1. Locate a cross section of the stream channel for discharge determination that has most of the following qualities:
   - Segment of stream above and below cross section is straight.
   - Depths mostly greater than 15 cm and velocities mostly greater than 0.15 m/s. Do not measure discharge in a pool.
   - U shape, with a uniform streambed free of large boulders, woody debris or brush, and dense aquatic vegetation.
   - Flow is relatively uniform, with no eddies, backwaters, or excessive turbulence.

2. Lay the surveyor's rod (or stretch a meter tape) across the stream perpendicular to its flow, with the "zero" end of the rod or tape on the left bank, as viewed when looking downstream. Leave the tape tightly suspended across the stream, at the bankfull[a] mark or higher. Adjust the tape with the aid of a small bubble level suspended from the rod or tape so it is, and remains throughout the period of measurement, level.

3. Attach the velocity meter probe to the calibrated wading rod. Check to ensure the meter is functioning properly and the correct calibration value is displayed. Calibrate (or check the calibration) the velocity meter and probe as directed in the meter's operating manual. Place an X in the VELOCITY AREA box on the Stream Discharge form.

4. Divide the total wetted stream width into 15 to 20 equal-size intervals. To determine interval width, divide the width by 20 and round up to a convenient number. Intervals should not be less than 10 cm wide, even if this results in less than 15 intervals. The first interval is located at the left margin of the stream (left when looking downstream), and the last interval is located at the right margin of the stream (right when looking downstream).

5. Stand downstream of the rod or tape and to the side of the first interval point (closest to the left bank if looking downstream).

6. Place the wading rod in the stream at the interval point and adjust the probe or propeller so that it is at the water surface. Place an X in the appropriate DISTANCE UNITS and DEPTH UNITS boxes on the Stream Discharge form. Record the distance from the left bank and the depth indicated on the wading rod on the Stream Discharge form.

   **Note:** for the first interval, the distance equals 0 cm, and in many cases depth may also equal 0 cm. For the last interval, the distance will equal the wetted width (in cm), and depth may again equal 0 cm.

*Continued*

**TABLE B.5 (*Continued*)   VELOCITY-AREA PROCEDURE FOR DETERMINING STREAM DISCHARGE**

7. Stand downstream of the probe or propeller to avoid disrupting the stream flow. Adjust the position of the probe on the wading rod so it is at 0.6 of the measured stream depth below the surface of the water. Face the probe upstream at a right angle to the cross section, even if local flow eddies hit at oblique angles to the cross section.

8. Wait 20 s to allow the meter to equilibrate, then measure the velocity. Place an *X* in the appropriate VELOCITY UNITS box on the Stream Discharge form. Record the value on the Stream Discharge form. Note for the first interval, velocity may equal 0 because depth will equal 0. Note that negative velocity readings are possible; when recording negative values, assign a flag to denote they are indeed negative values.
   • *For the electromagnetic current meter (e.g., Marsh-McBirney)*, use the lowest time constant scale setting on the meter that provides stable readings.
   • *For the impeller-type meter (e.g., Swoffer 2100)*, set the control knob at the midposition of DISPLAY AVERAGING. Press *RESET* then *START* and proceed with the measurements.

9. Move to the next interval point and repeat steps 6 through 8. Continue until depth and velocity measurements have been recorded for all intervals. Note that for the last interval (at the right margin), depth and velocity values may equal 0.

10. At the last interval (the right margin), record a *Z* in the FLAG field on the field form to denote the last interval sampled.

ª Physical indicators of the bankfull stage include (1) top of highest depositional features, (2) break in the slope of the bank or a change in particle size, (3) staining of rocks, and (4) exposed root hairs below an intact soil.

known volume. The channel is considered to be small if it is so shallow that the current velocity probe cannot be placed in the water or, where the channel is broken up and irregular because of rocks and debris, a suitable cross section for using the velocity-area procedure is not available. The timed-filling method can be extremely precise and accurate but requires a natural or constructed spillway of free-falling water. Because obtaining data by this procedure can result in channel disturbance or stirring up a lot of sediment, wait until after all biological and chemical measurements and sampling activities have been completed before measuring discharge.

It can be helpful if you choose a cross section of the stream that contains one or more natural spillways or plunges that collectively include the entire stream flow. You can measure discharge at each location and add the measurements together. A temporary spillway can be constructed using a portable V-notch weir, plastic sheeting, or other materials (i.e., rocks, wood) that are carried with you or are available on the site. Choose a location within the sampling reach that

## TABLE B.6    TIMED-FILLING PROCEDURE FOR DETERMINING STREAM DISCHARGE

*Note*: If measuring discharge by this procedure will result in significant channel disturbance or will stir up sediment, do not determine discharge until all biological and chemical measurement and sampling activities have been completed.

1. Choose a cross section that contains one or more natural spillways or plunges, construct a temporary spillway using on-site materials, or install a portable weir using a plastic sheet and on-site materials.

2. Place an *X* in the TIMED FILLING box in the stream discharge section of the Stream Discharge form.

3. Position a calibrated bucket or other container beneath the spillway to capture the entire flow. Use a stopwatch to determine the time required to collect a known volume of water. Record the volume collected (in liters) and the time required (in seconds) on the Stream Discharge form.

4. Repeat step 3 a total of five times for each spillway that occurs in the cross section. If there is more than one spillway in a cross section, you must use the timed-filling approach on all of them. Additional spillways may require additional data forms.

is narrow and easy to block when using a portable weir. Position the weir or constructed spillway in the channel so that the entire flow of the stream is completely rerouted through its notch. Impound the flow with the weir, making sure that water is not flowing beneath or around the sides of the weir. Use mud or stones and plastic sheeting to obtain a good waterproof seal. The notch must be high enough to create a small spillway as water flows over its sharp crest.

The timed-filling procedure is outlined in Table B.6. Make sure that the entire flow of the spillway goes into the bucket. Record the time it takes to fill a measured volume on the Stream Discharge section of the Stream-Water-Sampling Record Form. Repeat the procedure five times. Discharge will be calculated as an average of these five measurements. If the cross section contains multiple spillways, you will need to do separate determinations for each spillway. If so, clearly indicate which time and volume data replicates should be averaged together for each spillway; use an additional Stream-Water-Sampling Record Form if necessary. On the additional form, record a flag value (e.g., F1) on all lines in the Timed Filling section and explain in the comment section that the flag means an additional spillway was measured.

## B.5.4  Equipment and Supplies

Table B.7 shows the list of equipment and supplies necessary to measure stream discharge. Use this checklist to ensure that equipment and supplies are organized and available at the stream site so the activities are conducted efficiently.

**TABLE B.7    EQUIPMENT AND SUPPLY CHECKLIST FOR MEASURING STREAM DISCHARGE**

| Quantity | Item | ✓[a] |
|---|---|---|
| 1 | Surveyor's telescoping leveling rod (7-m long, metric scale, round cross section) | |
| 1 | 50-m fiberglass measuring tape and reel | |
| 1 | Small bubble level to make sure the tape is level | |
| 1 | Current velocity meter, probe, and operating manual | |
| 1–2 | Extra batteries for velocity meter | |
| 1 | Top-set wading rod (metric or English scale) for use with current velocity meter | |
| 1 | Portable weir with 60/V notch (optional) | |
| 1 | Plastic sheeting to use with weir (optional) | |
| 1 | Plastic bucket (or similar container) with volume graduations | |
| 1 | Stopwatch | |
| 1 | Covered clipboard | |
| | Soft (no. 2) pencils | |
| | Stream Discharge forms (one per stream plus extras if needed for timed-filling procedure or additional velocity-area intervals) | |
| 1 copy | Field operations and methods documents | |
| 1 set | Laminated sheets of procedure tables or quick reference guides for stream discharge | |

[a] Check off each item as it is added.

## B.6   ACKNOWLEDGMENTS

This SOP is based partly on material developed by Kaufmann (2006), Herlihy (2006), and Merritt et al. (1997).

## REFERENCES

Baker, J.R., D.V. Peck, and D.W. Sutton (Eds.). 1997. *Environmental Monitoring and Assessment Program Surface Waters: Field Operations Manual for Lakes.* Report No. EPA/620/R-97/001. US Environmental Protection Agency, Washington, DC.

Herlihy, A.T. 2006. Section 5. Water chemistry. In D.V. Peck, A.T. Herlihy, B.H. Hill, R.M. Hughes, P.R. Kaufmann, D.J. Klemm, J.M. Lazorchak, F.H. McCormick, S.A. Peterson, P.L. Ringold, T. Magee, and M. Cappaert. *Environmental Monitoring and Assessment*

Program—*Surface Waters Western Pilot Study: Field Operations Manual for Wadeable Streams.* EPA/620/R-06/003. US Environmental Protection Agency, Office of Research and Development, Washington, DC, pp. 85–98.

Kaufmann, P.R. 2006. Section 6. Stream discharge. In D.V. Peck, A.T. Herlihy, B.H. Hill, R.M. Hughes, P.R. Kaufmann, D.J. Klemm, J.M. Lazorchak, F.H. McCormick, S.A. Peterson, P.L. Ringold, T. Magee, and M. Cappaert. *Environmental Monitoring and Assessment Program—Surface Waters Western Pilot Study: Field Operations Manual for Wadeable Streams.* EPA/620/R-06/003. US Environmental Protection Agency, Office of Research and Development, Washington, DC, pp. 99–110.

Linsley, R.K., M.A. Kohler, and J.L.H. Paulhus. 1982. *Hydrology for Engineers.* McGraw-Hill, New York.

Merritt, G.D., V.C. Rogers, and D.V. Peck. 1997. Section 3. Base site activities. In J.R. Baker, D.V. Peck, and D.W. Sutton (Eds.), *Environmental Monitoring and Assessment Program Surface Waters: Field Operations Manual for Lakes.* Report No. EPA/620/R-97/001. US Environmental Protection Agency, Washington, DC, pp. 3-1–3-14.

Rantz, S.E. and others. 1982. *Measurement and Computation of Streamflow: Volume 1. Measurement of Stage and Discharge.* US Geological Survey Water-Supply Paper 2175. US Government Printing Office, Washington, DC. http://pubs.usgs.gov/wsp/wsp2175/pdf/WSP2175_vol1a.pdf.

# Appendix C: Basic Standard Operating Procedures for Lake Field-Sampling Activities

## CONTENTS

The purpose of this appendix is to provide basic standard operating procedures (SOPs) for field sampling focused on measurement of lake chemistry as influenced by atmospheric deposition. The recommended protocol featured in the main body of this book provides guidelines regarding how to implement a field-sampling program for water chemistry, with explanation of some of the reasons why certain steps and precautions are recommended. An SOP is a detailed explanation of sequential steps to be taken in carrying out the water sampling. The SOP is based on the principles outlined in the protocol. In some cases, there are multiple "correct" ways to carry out a component of the field sampling. The basic principles remain the same, and these are reflected in the protocol. Nevertheless, the exact steps may differ and yet still satisfy the aims of the protocol. Thus, these basic SOPs may be modified by a particular sampling program or field office, as appropriate to local conditions. In modifying aspects of the SOPs, the guidelines represented in the protocol should always be carefully considered.

## C.1 INTRODUCTION

This SOP provides guidelines for lake sampling. It is intended as a base SOP, suitable for adoption as a stand-alone procedure or for modification to fit local program needs. It is divided into individual sections that cover pretrip activities, sampling-site documentation, index site location, lake sampling, lake assessment, post-trip activities, and needed equipment and supplies.

## C.2 PRETRIP ACTIVITIES

### C.2.1 Background

Field teams conduct a number of activities in their office or at a base site prior to departing for the field sampling. These include tasks that must be completed both before departure to the sampling site and after return from the site. This section describes procedures for office and base site pretrip activities that should be carried out in support of lake sampling.

Predeparture activities include development of sampling itineraries, instrument calibration if appropriate, equipment checks and repair, generation of supply inventories, and sample container preparation. Procedures for these activities are described in the following sections.

Before leaving the base location, package the sample containers (typically two sample bottles and two 60-ml syringes or glass bottles with septum caps) for each site to be sampled (plus backup). Fill out a set of water chemistry sample labels. Attach a completed label to each sample bottle and syringe. Make sure the syringe labels do not cover the volume gradations on the syringe.

## C.2.2   Daily Itineraries

Field-sampling efforts should include a project leader who guides activities in the field and a project coordinator who remains in the office during the sampling effort. The project leader reviews each site folder to ensure that it contains the appropriate maps, contact information, copies of access permission letters (if needed), and access instructions. Additional activities can include confirming the best access routes, calling landowners or local contacts (if applicable), confirming lodging or camping plans and locations (with directions), and coordinating rendezvous locations with individuals who must meet with field teams prior to accessing a site. This information is used to develop an itinerary. The project leader should provide the project coordinator with a schedule for each day of sampling. Schedules include departure time, estimated duration of sampling activities, routes of travel, and estimated time of arrival at the sampling sites and return to the base site. Changes that might be made to the itinerary should be relayed by the project leader to the project coordinator as soon as possible. Miscommunications can result in the initiation of expensive search-and-rescue procedures and disruption of carefully planned schedules.

## C.2.3   Instrument Checks and Calibration

If appropriate, each field team should test or calibrate field instruments prior to departure for the sampling site. Such testing may be appropriate for dissolved oxygen meters, global positioning systems (GPSs), and perhaps other instrumentation. Batteries should be checked prior to departure for field sites. Extra batteries should be carried.

Field personnel should check the inventory of supplies and equipment prior to departure using site visit checklists. Meters, probes, and sampling gear should be packed for transport to the field in such a way that it minimizes physical shock and vibration during transport. Rafts or float tubes should be packed for transport to minimize the potential for puncture by any sharp object.

## C.2.4   Supply Inventories

Develop a checklist of equipment and supplies that will be needed to conduct lake sampling. Check off each item as it is packed and loaded for transport to the field.

A preliminary list of equipment and supplies required to collect lake samples and associated field data is presented in Table C.1. Use and revise this checklist to ensure that equipment and supplies are organized and available at the lake-sampling site in order to conduct the activities efficiently.

## C.2.5   Sample Container Preparation

Generally, it is the responsibility of the analytical laboratory to provide the field team with properly washed bottles and syringes as needed to carry out the sampling program. To do this, the laboratory will need to know in advance what analytes will be measured in the samples to be collected. Ensure that the proper number, type, and size sampling containers are provided. It is wise to carry a few backups.

## C.3   SITE DOCUMENTATION

### C.3.1   Background

Here, we describe SOPs for establishing and documenting sampling sites on small-to-medium-size lakes, primarily those situated in relatively remote backcountry locations. This procedure applies to new sites for which approximate locations have been designated based on program objectives and sampling design. It also applies to previously established sites for which current or updated site documentation is needed.

Sampling the correct lake is critical to most lake study sampling designs. It is also important to identify, to the extent possible, the index site (deepest point) on a lake. On arriving at a target lake, the GPS is a valuable tool to verify identity and location. Nevertheless, site verification must be supported by all available information (e.g., maps, road signs, GPS, and expected lake size and shape). Do not sample the lake if there is reason to believe it is the wrong one. Contact the project coordinator to resolve discrepancies.

**TABLE C.1  CHECKLIST OF EQUIPMENT AND SUPPLIES FOR SAMPLING WATER CHEMISTRY AND SECCHI DEPTH**

| Quantity | Item |
|---|---|
| | **Standard Items** |
| 1 | Field thermometer |
| 1–2 | Sample bottle(s) with completed sample label attached (in clean plastic bag); Include second bottle if sampling at that site is to be replicated |
| 2–4 | 60-ml plastic syringes (with Luer-type tip) or glass bottles with septum caps with completed sample labels attached |
| 1 | Plastic container with snap-on lid to hold filled syringes |
| 2–4 | Syringe valves (Mininert with Luer-type adapter, or equivalent, available from a chromatography supply company) |
| 1 | Cooler with 4 to 6 plastic bags (1-gal) of ice **or** a medium or large opaque garbage bag to store the water sample at shoreline |
| 1 | Lake-Water-Sampling Record form |
| 1 set | Water chemistry labels (if not filled out and attached at base site) |
| 2–4 | Soft-lead pencils and write-in-rain-type pens for filling out field data forms and notebook entries |
| 2–4 | Fine-tip indelible waterproof markers for filling out labels |
| 1 copy | Field operations and methods documents |
| 2–4 | Plastic gloves stored in a secure plastic bag |
| 1 | Survey-grade global positioning system and compass |
| 1 | Digital camera with extra memory cards and batteries |
| 1 | Backpack with waterproof cover (if site is not accessible by vehicle) |
| 1 | Van Dorn sampler with messenger and cable |
| 1 | Raft or float tube with pump for inflating |
| 1 | First aid kit |
| 1 | Locally determined safety equipment |
| 1 | Secchi disk and line (with depth increments) |
| 1 | Tape measure |
| | **Optional Items** |
| Roll or box of tape strips | Clear packaging tape to cover labels |
| 1 | Dissolved oxygen/temperature meter with probe |

**TABLE C.1 (*Continued*)   CHECKLIST OF EQUIPMENT AND SUPPLIES FOR SAMPLING WATER CHEMISTRY AND SECCHI DEPTH**

| Quantity | Item |
|---|---|
| 1 | DO repair kit containing additional membranes and probe-filling solution |
| 1 | Conductivity meter with probe |
| 1 | 250- or 500-ml plastic bottle of conductivity quality control check sample (QCCS) labeled *RINSE* (in plastic bag) |
| 1 | 250- or 500-ml plastic bottle of conductivity QCCS labeled *TEST* (in plastic bag) |

## C.3.2   Objective

The objective of this procedure is to establish and document new lake-sampling sites and to update documentation for established sites, providing

- Site descriptions and notes
- Travel and access descriptions and notes
- Site coordinates obtained in the field using a GPS unit
- Site and access-related photos

For established sites, existing site documentation will be evaluated for clarity and improved as needed based on conditions observed in the field.

## C.3.3   Material Needed for Use in the Field for Site Documentation

A. Available site documentation records for previously established sites:
   1. Site location maps, topographic maps, and road maps
   2. Site descriptions and access notes
   3. Site coordinates
   4. Site photos
B. Preliminary site documentation for new sites:
   1. Site location maps, topographic maps, and road maps, indicating approximate site locations
   2. General site descriptions and access notes
C. General material for site documentation:
   1. Regional-scale topographic and road maps
   2. Lake-Sampling-Site Documentation Forms on waterproof paper
   3. Clipboard or field notebook and pens for use with waterproof paper
   4. GPS unit with replacement batteries

5. Digital camera with charged battery and charged replacement battery
6. Gate keys (if needed)
7. Cell phone with numbers of project staff and management agency offices

## C.3.4 Sequence of Site Documentation Activities

A. Initiate the Lake-Sampling-Site Documentation Form prior to entering the field. Enter the official USGS and local lake names and ID, forest or wilderness, and the name, affiliation, e-mail address, and phone number of the person responsible for site documentation.
B. Fill in requested information on the date and visit number, the approximate area of the lake (estimated visually to confirm that it generally agrees with what is seen on the topographic map), and access information.
C. Select or locate, using GPS, the specific sampling site (applies to new sites).
   1. The approximate or preliminary location of new site locations will be indicated on topographic maps. The sample collection team must still determine the exact point on the lake to be sampled.
   2. The preferred sampling-site location is over the deepest portion of the lake, which is often, but not always, near midlake. This is designated as the "index" site. If it is not feasible to access the index site or for conducting some types of screening studies, it is acceptable to sample from the principal outlet stream (designated "outlet" sample) rather than at the index site. If there is no available outlet stream or if the outlet stream is not flowing at a sufficient rate to collect a representative sample, then it can be acceptable to collect a sample by reaching into the lake from an appropriate location along the lakeshore (designated "shoreline" sample).
D. Obtain new coordinates at the sampling site or verify the correct location using the GPS.
   1. The unit position format should be set to decimal degrees (hddd.ddddd). The datum should be set to NAD83. Resolution should be expressed in meters.
   2. When "Mark Waypoint" is selected, the default GPS site ID (a number) should be changed to the actual site ID.
   3. Before saving the coordinates, note the estimated accuracy of the measurement and enter on the Sampling-Site Documentation Form.
   4. Save the coordinates on the GPS and record the coordinates in decimal degrees, the datum and the elevation (preferably meters) on the Sampling-Site Documentation Form. Use NAD83 if possible to

conform to National Resource Information System (NRIS) requirements. Do not rely solely on the GPS to store the coordinates.

5. Confirm that the waypoint has been saved in the GPS unit.

6. It can be helpful to establish and document benchmarks on shoreline rocks.

E. Enter (if applicable) the approximate lake water level on the Sampling-
Site Documentation Form.

F. Enter site description information on the Sampling-Site Documentation
Form.

1. For existing sites, enter information to improve and update existing site description information.

2. Generally describe the site, referring to proximity to landmarks
(trails, bridges, tributaries, trees, shoreline features, landscape features, or other relatively permanent features). Add any information
that will help future sample collectors indentify the site. Also, add
any information here that might be relevant to water quality, such
as cleared land, mining (ongoing or historical), roads, construction, logging, development, or any earth disturbance observed on
or near the shoreline, above the lake in the watershed, or along the
inlet streams.

G. Enter travel and access directions on the Sampling-Site Documentation
Form.

1. For existing sites, improve and update existing travel and access
information.

2. Travel and access notes should be sufficient to guide future sample
collectors to the site without reliance on GPS units. Not all future
sample collectors will necessarily have GPS units.

3. Access notes should refer to trails, roads, and permanent landmarks, providing distances and, where helpful, compass-based
directions. Backtrack if necessary to determine distances. Linear
distances and directions from the established site waypoint can
be determined using the GPS unit.

4. If a parking location is not immediately adjacent to the sampling
site, use the GPS unit to obtain the coordinates for the parking
location and record them in the travel and access information
entry area of the Sampling-Site Documentation Form.

5. For complicated or long walk-ins, use a GPS unit to record and
save a track. But, again, do not rely on future sample collectors
having access to a GPS unit.

6. Sketch the route on the back of the Sampling-Site Documentation
Form for scanning and saving as a JPEG image if that would be
helpful.

H. Make a sketch of the lake on the Sampling-Site Documentation Form. Mark on the sketch the launch and sampling-site locations.

I. Obtain site and access photos.

   1. Photos should be obtained providing views of the sampling site and the shoreline and views of other distinguishing features in relation to the site (bridges, roads, notable rocks, trees, landforms, signage, etc.). Photos should also be obtained to show important aspects of site access (parking area, forks in the trail, etc.).

   2. All photos should be listed on the form, including the file name, date, and description. Enter this information at the time that the pictures are taken. Do not rely on memory for later entry of photo descriptions. The entered description should serve as the photo caption for site documentation reporting.

## C.3.5 Lake Verification at the Launch Site

Record directions to the sampling site, and a description of the launch location for lake sampling, on the Lake-Sampling-Site Documentation Form in the site information folder. This information will be important in the future if the site is revisited by another sampling team. Provide information about signs, road numbers, gates, landmarks, and any additional information you feel will be useful to another sampling team in relocating this site. It is also helpful to describe the road distance traveled (miles) between turns and hiking distance or time traveled to reach the sampling or launch site. Additional details can also be helpful. What landmarks are in the vicinity of the site? Is the trailhead well marked?

If a GPS fix is obtained, record the location in decimal degrees and the type of satellite fix (two- or three-dimensional) for the site. Compare the site information folder map coordinates recorded for the site with the GPS coordinates displayed at the site. Check to see if the two sets of coordinates are within a distance that is approximately equal to the precision of the GPS receiver without differential correction of the position fix. If a GPS fix is not available, do not record any information but try to obtain the information at a later time during the visit. A fix may be taken at any time during a site visit and recorded on the form. If this is the first visit to this lake, mark the location of the launch site with an L on the lake outline that is provided on the Lake-Sampling-Site Documentation Form. In addition to the GPS, use as many of the following methods as possible to verify the site:

1. Obtain confirmation from a local person familiar with the area.
2. Identify confirming trails, roads, and signs.
3. Compare lake shape to that shown on the topographic map included in the site information folder.

4. Determine lake position relative to identifiable topographic features shown on the map.
5. Compare visual evaluation of lake area with available mapped information.

If this is not the first visit to this lake and if the lake shape on the map sketch that appears on the Lake-Sampling-Site Documentation Form and on the US Geological Survey (USGS) map do not correspond with each other or with the actual lake shape as seen in the field, check "Not Verified" and provide comments on the form. The lake should *not* be sampled if there are clear major differences in lake shape or lake area. At each lake, evaluate whether the lake meets the standard definitions of what constitutes a lake, for example, at least:

- 1 ha total surface area
- 100 m$^2$ open water
- 1 m in depth

Depending on the scope of the particular study, if the lake does not fit this definition, it may be appropriate to check "nontarget" in the lake-sampled section on the bottom of the Lake-Sampling-Site Documentation Form and provide an explanation for not sampling the lake. Add any comments as appropriate.

## C.3.6  Index Site Location

Locate the sampling site in what is approximately the deepest portion of the lake. There are different ways to do this, as follows:

1. If the deepest location had been determined and documented on a previous trip to this lake, based on that documentation and use of GPS or mapped lake features, navigate to the sampling location.
2. If the sampling location has not previously been documented, locate the deepest part of the lake based on visual examination of the lake shape and surrounding topography, coupled with reconnaissance on foot or by boat for up to about one-half hour. Use visual cues or soundings with a weighted line to locate what appears to be the deepest part of the lake.

Once the sampling location has been selected, at what appears to be the deepest part of the lake, determine the GPS coordinates and record them on the Lake-Sampling-Site Documentation Form. Mark the sample site with an *x* in the lake drawing. A checklist for lake verification is given in Figure C.1.

| LAKE VERIFICATION CHECKLIST | |
|---|---|
| ✓ | |
| | Site information folder for lake to be sampled |
| | Clipboard |
| | Lake Sampling Site Documentation Form |
| | Field notebook |
| | Sampling permit (if needed) |
| | GPS unit with manual, extra battery pack |
| | 50 m line to attach to rock anchor |

**Figure C.1**    Lake verification checklist.

## C.4  LAKE SAMPLING

### C.4.1  Background

These procedures cover collection of lake water samples and measurement of Secchi depth (transparency). The lake-sampling procedures assume collection of the primary sample from the deepest part of the lake. Measurement of Secchi depth and collection of the deep-water index sample will require use of a boat or float tube. If it is not possible to sample the lake by boat or float tube, the next-best option is to sample at the principal outlet stream. If a lake outlet sample is to be collected, instead of a sample in deep water, follow the procedures outlined in the stream-sampling SOP and sample the outlet stream as close to the lake as is practical.

If neither a deep-water sample nor an outlet sample can be collected, the third option is to sample from the shoreline, satisfying as many of the following criteria as possible:

- as close to the outlet as possible
- from a bedrock outcropping or otherwise rocky area
- from the deepest accessible point

Water must be deep enough so that surface scum and sediments are not collected into the bottle, preferably in a wind-exposed area so that the water is relatively well mixed. Avoid sampling in locations having emergent vegetation or downed logs or other woody debris. Avoid skinny-dipping in the lake immediately before you sample it.

If desired, *in situ* measurement of one or more parameters can be made using a multiparameter water quality sonde. Such measurements might include temperature, pH, dissolved oxygen, specific conductance, or turbidity. The procedures for such *in situ* data collection will vary with the specific field instrument, but in general require the following steps:

- Initiate water quality sonde field calibration and calibration checks. Record results on a water quality instrument calibration and post-calibration record form. Confirm that sensor check criteria are met. If criteria are not met, recalibrate, perform sensor maintenance, or replace sonde or sensors as needed to meet the criteria.
- Deploy the water quality sonde for the period required to obtain stabilization. Enter the results and time of measurement in the *in situ* Water Quality Data section of the Lake-Water-Sampling Record Form.

## C.4.2  Documentation of Data and Sample Collection

The Lake-Water-Sampling Record Form is used to document sample collection and field data. The form is used to record the following information:

A. The station ID and station name.
B. The date and arrival time for the site visit and specific times of measurements obtained.
C. The name, contact information, and affiliation of the individual who is the collector of record and responsible for protocol adherence during the site visit.
D. Suggested revisions or amendments to site documentation and travel directions.
E. A listing of site-related photographs taken, including file name, date, and descriptions.
F. Qualitative descriptions of weather, lake level and appearance, and other factors that might influence water quality during the site visit.
G. Air temperature.
H. Results for all water quality data collected *in situ*, including
    1. Numerical results, units, and measurement time.
    2. Instruments used and methods identification.
I. Identification of calibration and postcalibration sensor check records.
J. A listing of all samples collected, including
    1. Collection time.
    2. Types of samples collected and number of replicates.

## C.4.3   Sequence of Activities for Data Collection

The typical sequence of activities on arrival at the lake-sampling site is as follows:

A. Confirm the site location based on information in the Lake-Sampling-Site Documentation Form, including coordinates, photos, and access notes.
B. Initiate completion of the Lake-Water-Sampling Record Form.
C. Complete Site Information and General Observations sections of the Lake-Water-Sampling Record Form.
D. Enter information needed to improve or correct the site description and travel directions provided on the Lake-Sampling-Site Documentation Form.
E. Obtain any photographs needed to improve site documentation and enter file names, dates, and descriptions.
F. Note any factors (other than weather and lake level) that might affect water quality (e.g., shoreline or watershed disturbance, debris in water).
G. Collect water samples and complete the Water Sample section of the Lake-Water-Sampling Record Form. Enter any *in situ* data into that section of the Lake-Water-Sampling Record Form.
H. Complete the Chain-of-Custody Form.
I. Check to make sure that all of the information recorded on the sample labels, Chain-of-Custody Form, and Lake-Sampling Record Form match.
J. Obtain lake-level measurements, if required.

## C.4.4   Sample Collection

### C.4.4.1   Deep-Water Index Sample

Collect a water sample at the index site using a Van Dorn water sampler from 1.5-m depth (0.5 m if lake depth is less than 2.0 m), using the procedure described in Table C.2. From the Van Dorn sampler, fill the required number of syringes or glass bottles and one or two 500- or 1000-ml sample bottles. Procedures for collecting these samples are presented in Table C.3. Prior to filling syringes and sample bottles, check the labels on these containers to ensure that all written information is legible and that each container has the same (and the correct) site identification number. It can be a good idea also to place clear packing tape over the label and identification code, covering the label completely. Record the identification code assigned to the sample set (the syringes and bottles collected from the same site are considered one sample) on the Lake-Water-Sampling Record Form. Also, record the depth from which the sample was collected (usually 1.5 or 0.5 m) on the form. Enter a flag code and provide comments on the Sample Collection form if there are any problems in collecting the sample or if conditions occur that may affect sample integrity. Store samples in the appropriate containers in the dark

## TABLE C.2    SAMPLE COLLECTION USING VAN DORN SAMPLER

1. Open the Van Dorn sampler by pulling the elastic bands and cups back and securing the latches. Make sure that the mechanism is cocked so that it will be tripped by the messenger weight. Make sure that all valves are closed. Inspect the line for fraying, especially where it connects to the Van Dorn. **Do not place hands inside or on the lip of the container; this could contaminate samples. To reduce chances of contamination, wear powder-free latex laboratory gloves.**

2. Attach the free end of the messenger line to the boat. This is important to prevent accidental loss of the equipment overboard. Rinse the open sampler by immersing it in the water column three times.

3. Lower the sampler to 1.5 m below the surface (0.5 m in lakes < 2-m deep).

4. Trip the sampler by releasing the messenger weight so that it slides down the line.

5. Raise the full sampler out of the lake. Set it on a clean, flat surface in an upright position. To avoid contamination, do not set the sampler in the bottom of the boat. Applying some body weight to the top of the Van Dorn sampler often will seal minor air leaks and preserve the sample integrity. If air enters the Van Dorn sampler, discard the sample and obtain another (repeat steps 1–5).

## TABLE C.3    SYRINGE AND SAMPLE BOTTLE COLLECTION[a]

1. Make sure that the sample bottle(s) and 60-ml syringes have the same site identification code number (which identifies a single lake) and that the labels are completely covered with clear tape. Record the identification code number on the Sample Collection form.

2. Fill one prerinsed bottle for each routine lake water sample. Fill a second bottle, with its own unique sample ID/bar code, if this site is to be replicated for quality assurance/quality control (QA/QC) purposes.

3. Unscrew the valve at the top of the Van Dorn sampler. Fit a prelabeled syringe to the fitting.

4. Slowly withdraw a 20-ml aliquot into the 60-ml prelabeled syringe. Pull the plunger back so that the water contacts all inner surfaces of the syringe. Expel the water from the syringe. Repeat this rinse procedure twice more (there are three rinses for each syringe sample).

5. Reattach the syringe to the Van Dorn sampler and slowly withdraw 60 ml of water into the syringe. If air enters the Van Dorn sampler during this process, dispose of the sample and obtain another Van Dorn sample.

6. Place the syringe valve on the syringe tip. Press the green button toward the syringe.

*Continued*

**TABLE C.3 (*Continued*)   SYRINGE AND SAMPLE BOTTLE COLLECTION[a]**

7. Hold the syringe with the tip and valve pointed skyward. Tap the syringe to gather air bubbles to the top. Expel all air from the syringe and press the red button on the syringe valve to seal the syringe with 40 to 50 ml of sample water remaining.

8. Repeat steps 2 to 5 for one to three additional syringes. There should be a total of two syringes for each routine water sample (four syringes if sample is being replicated).

9. Place the syringes in the solid plastic container and place in the cooler. Use ice contained in sealed 1-gal plastic bags to maintain the sample below 4°C.

10. Unscrew the top valve of the Van Dorn sampler. Unscrew the lid of the prelabeled sample bottle.

11. Open the bottom valve of the Van Dorn sampler and partially fill the sample bottle with water (approximately 50 ml).

12. Screw the lid on the bottle. Shake the bottle so that the water inside contacts all sides. Discard the water. Repeat this rinse procedure twice more. Collection of the water sample in the bottle should be preceded by three rinses.

13. Open the Van Dorn valve and completely fill the bottle.

14. Compress the plastic bottle to remove any residual headspace. Seal the cap tightly. Holding the glass bottle (if applicable) level, fill it completely to the top. Seal the cap tightly.

15. Place bottle in a cooler with sealed 1-gal plastic bags of ice. Note the depth from which the sample was collected on the Sample Collection form.

*Source:* Modified from Baker, J.R. and D.V. Peck. 1997. Section 4. Lake verification and index site location. In J.R. Baker, D.V. Peck, and D.W. Sutton (Eds.), *Environmental Monitoring and Assessment Program Surface Waters: Field Operations Manual for Lakes.* Report No. EPA/620/R-97/001. US Environmental Protection Agency, Washington, DC.

[a] Wear powder-free surgical gloves while collecting syringe and bottle samples. Syringes may be chilled before use to reduce the occurrence of air bubbles in the sample.

and verify that they are carefully packed with plenty of ice bags and properly positioned, sealed, and labeled in the sample coolers. Recheck all forms and labels for completeness.

## C.4.4.2  Lake Outlet Sample
To collect a lake outlet sample, follow the procedures outlined in the stream-sampling SOP. Collect the sample from the outlet stream as close to the lake as is practical.

### C.4.4.3  Shoreline Sample

Only collect a shoreline sample if the study objective is to perform a rough screening to identify probable lake chemical conditions or if it is not feasible to collect either a deep-water or outlet sample from the subject lake. Collect the shoreline sample as follows: In the field, make sure that the labels all have the sample ID number (bar code) and that the labels on the bottles and syringes are securely attached. Carefully avoid disturbance of water or sediment in the vicinity of the sampling point prior to sample collection. This means not walking in the water or on loose rocks. If you must walk out to obtain a clean sample, wait for the sediment to settle before collecting the sample.

Collect a water chemistry sample, as described in Table C.4, in the deepest water possible. Throughout the collection process, it is important to take precautions to avoid contaminating the sample. Rinse all sample containers three times with lake water before filling them with the sample. Many remote lakes have a low ionic strength and can be contaminated easily by perspiration from hands, sneezing, smoking, insect repellent, sunscreen, or chemicals (e.g., formalin or ethanol) used when collecting other types of samples. Make sure that none of the water sample contacts your hands before going into the sample bottle or syringe. The chemical analyses conducted using the syringe or septum bottle samples can be affected by equilibration with atmospheric carbon dioxide; thus, it is essential that no outside air contact the syringe samples during or after collection.

The sample should be collected on a step-by-step basis as follows:

a. Remove the gloves from the plastic bag and put them on.
b. Remove the sample bottle from the plastic bag. Do not put the bag on the ground.
c. Rinse the sample bottle in the lake at a location at least 10 feet away from the sample collection point. The bottle and cap should be rinsed three times. For each rinse, fill the bottle and then pour the rinse water over the inside of the cap, held bottom-side up in the other hand. Pour the rinse water away from the lake-sampling point and avoid stirring up lake bed debris during the process.
d. After the rinsing is completed, move to the sampling point and collect the sample by submersing the tilted bottle or syringe to a depth midway between the sediment and the water surface. Fill the bottle as completely as possible. While collecting the sample, avoid stirring up lake bed debris that might be collected with the sample. If it is deemed that debris may have entered the sample bottle, discard the contents (at a different location), rerinse the bottle (or use a clean backup bottle), and collect a new sample.

**TABLE C.4    SHORELINE SAMPLE COLLECTION PROCEDURES FOR WATER CHEMISTRY**

### Collection into Bottle

1. Rinse the sample bottle and lid three times with stream water. Discard the rinse downstream.

2. Check to ensure that the correct labels are affixed to each sample bottle and syringe. Fill the sample bottle(s), holding the bottle in a tilted position approximately at the midpoint between the water surface and the lake bed, being careful not to disturb any sediment prior to or while collecting the sample. Try to avoid generating large bubbles in the bottle while it is being filled. If a septum cap is being used, place the cap on the bottle under the surface of the water to avoid any contact of the sample with the air.

3. Place the sample bottle(s) in a cooler (on ice or stream water) and shut the lid. If a cooler is not available, place the bottle(s) in an opaque garbage bag and immerse it in the stream.

### Collection into Syringe

4. Submerge a 60-ml syringe halfway into the lake and withdraw a 15- to 20-ml aliquot. Pull the plunger to its maximum extension and shake the syringe so the water contacts all surfaces. Point the syringe away from the lake and discard the water by depressing the plunger. Repeat this rinsing procedure two more times.

5. Submerge the syringe into the lake again and slowly fill the syringe with a fresh sample. Try not to get any air bubbles in the syringe. If more than one or two tiny bubbles are present, discard the sample and draw another one.

6. Invert the syringe (tip pointing up) and cap it with a syringe valve. Tap the syringe lightly to detach any trapped air bubbles. With the valve open, expel the air bubbles and a small volume of water, leaving between 40 and 50 ml of sample in the syringe. Close the syringe valve. If any air bubbles were drawn into the syringe during this process, discard the sample and fill the syringe again (step 6).

7. Repeat steps 5 through 7 with a second syringe. Fill a total of four syringes if the lake sample is to be replicated. Place the syringes together in the cooler or temporarily (until time to depart from the lake) in the lake water with the sample bottle(s).

### Postsampling Actions

8. Record the sample ID number (bar code) on the Lake-Water-Sampling Record form along with the pertinent information (lake name, site ID, date, etc.). Note anything that could influence sample chemistry (heavy rain, potential contaminants) in the Comments section. If the sample was collected at the targeted site, record an $X$ in the correct field. If you had to move to another location to collect the sample, check the Other box and record detailed reasons or information in the Comments section.

**TABLE C.4 (*Continued*)    SHORELINE SAMPLE COLLECTION PROCEDURES FOR WATER CHEMISTRY**

9. Complete the Chain-of-Custody Form.
10. Check to make sure that all of the information recorded on the sample label(s), Chain-of-Custody Form, and Lake-Sampling Record Form match.
11. Place each filled bottle into a zipper-lock bag; place the filled syringes into a plastic box with snap-on lid. After carrying the samples to the vehicles, place the (bagged) bottle(s) and (boxed) syringes in a cooler and surround with 1-gal resealable plastic bags filled with ice. Double bag to avoid getting cooling water into sample bags.

e. Immediately after collecting the sample, place the lid on the bottle (tightly) and return the bottle to its plastic bag. Seal the bag.
f. Complete the Water Sample Collection section of the Lake-Water-Sampling Record Form while at the sample site.

Record the information from the sample label on the Lake-Water-Sampling Record Form. Note any problems related to possible contamination in the Comments section of the form.

a. Place the sample on ice or refrigerant immediately after collection. *Note*: Do not put ice in the plastic bag that contains the sample bottle.
b. For sites that are close to road access, the large sample cooler can be left in the collection team's vehicle. The samples can be placed in the cooler on return to the vehicle.
c. For sites that are not close to road access, the collection team should make arrangements to keep the samples on ice after collection and during the return hike. One approach would be to use a small soft-pack cooler or other container that will fit in a backpack. Ice, snow, or refrigerant could be placed in a small plastic bag in the cooler or container (double bag to avoid leakage and contamination of samples).
d. The samples should be kept in the dark and on ice until delivery to the lab. The ice may need to be replenished during sample transit. Avoid letting the sample bottle float in melted ice water. Do not place the sample bottle in a refrigerator or cooler with food or in any container that is not clean. Ship the samples to the laboratory as soon as possible, preferably within 24 h of sampling.

## C.4.5   Field Measurements

Anchor the boat if possible. After achieving a stable position and determining the site depth, measure Secchi disk transparency using the procedures in Table C.5. Record the depth of disk disappearance and the depth of

**TABLE C.5    SECCHI DISK TRANSPARENCY PROCEDURES**

1. Remove sunglasses unless they are prescription lenses.

2. Clip the calibrated chain (should already be in 0.5-m increments) to the Secchi disk. Make sure the chain is attached so that depth is determined from the upper surface of the disk.

3. Lower the Secchi disk over the shaded side of the boat until it disappears.[a]

4. Read the depth indicated on the chain. If the disappearance depth is less than 1.0 m, determine the depth to the nearest 0.01 m by marking the chain at the nearest depth marker and measuring the remaining length with a tape measure. Otherwise, estimate the disappearance depth to the nearest 0.1 m. Record the disappearance depth on the Lake-Water-Sampling Record form.

5. Slowly raise the disk until it reappears and record the reappearance depth on the form.

6. Note any conditions that might affect the accuracy of the measurement in the Comments field.

*Source:* Modified from Baker, J.R., and D.V. Peck. 1997. Section 4. Lake verification and index site location. In J.R. Baker, D.V. Peck, and D.W. Sutton (Eds.), *Environmental Monitoring and Assessment Program Surface Waters: Field Operations Manual for Lakes.* Report No. EPA/620/R-97/001. US Environmental Protection Agency, Washington, DC.

[a] If the disk is visible to the lake bottom, check the appropriate box on the form.

reappearance on the Lake-Water-Sampling Record Form. If the Secchi disk is visible at the bottom of the lake, check the Clear to Bottom box on the form. Comment on the form if there are any conditions that may affect this measurement (e.g., surface scum, suspended sediments, weather conditions).

Other field measurements might be made depending on the study. These could include measurements at the sample site or measurements at other locations or depths. They might include dissolved oxygen, specific conductance, or other parameters.

## C.5    GENERAL LAKE ASSESSMENT

### C.5.1    Background

Standard operating procedures are summarized here for the site assessment conducted at lake-sampling locations. The purpose of this assessment is to record site characteristics that may aid in the interpretation of the chemical or biological data collected from the lake.

## C.5.2  General Lake Assessment Procedures

Team members should complete the Lake Assessment portions of the Lake-Water-Sampling Record Form at the end of lake sampling, recording all observations from the lake that were noted during the course of the visit. This lake assessment is designed as a template for recording pertinent field observations. It is not intended to be comprehensive, and any additional observations should be recorded in the Comments section. The assessment consists of three major sections: General Lake Hydrologic Information, Shoreline Characteristics, and Qualitative Macrophyte Survey. Each is described next.

### C.5.2.1  General Lake Hydrologic Information

Observations regarding the general characteristics of the lake are described in Table C.6. The hydrologic lake type is an important variable for defining sub-populations for acidic deposition effects.

**TABLE C.6   GENERAL LAKE INFORMATION NOTED DURING LAKE ASSESSMENT**

| | |
|---|---|
| Hydrologic lake type | Note if there are any stream outlets from the lake, even if they are not flowing. If no lake outlets were observed, record the lake as a seepage lake. If the lake was created by a man-made dam (not that a dam is present just to raise the water level), record the lake as a reservoir. Otherwise, record the lake as a drainage lake. |
| Outlet dams | Note the presence of any dams (or other flow control structures) on the lake outlet(s). Differentiate between artificial (man-made) structures and natural structures (beaver dams). Describe in detail the observed flow control structure, providing measurements if possible. Note the material from which the structure is made. |
| Lake level | Examine the lake shoreline for evidence of lake-level changes (e.g., bathtub ring). If there are none, check "zero"; otherwise, try to estimate the extent of vertical changes in lake level from the present conditions based on other shoreline signs. |

*Source:* Herlihy, A.T. 1997. Section 9. Final lake activities. In J.R. Baker, D.V. Peck, and D.W. Sutton (Eds.), *Environmental Monitoring and Assessment Program Surface Waters: Field Operations Manual for Lakes*. Report No. EPA/620/R-97/001. US Environmental Protection Agency, Washington, DC.

## C.5.2.2   Shoreline Characteristics

Shoreline characteristics of interest during the lake assessment are described in Table C.7. To estimate the extent of major vegetation types, limit the assessment to the immediate lake shoreline area (i.e., within 20 m of the water). Also, estimate the percentage of the immediate shoreline that has been developed or modified by humans.

## C.5.2.3   Qualitative Macrophyte Survey

Macrophytes (aquatic plants large enough to be seen without magnification) can be important indicators of lake trophic (nutrient balance) status. The most important macrophyte indicator for assessment purposes is often the percentage of the lake area covered with macrophytes. For both "emergent/floating" and "submergent" coverage, choose one of the four percentage groupings (0–25%, 25–50%, 50–75%, 75–100%) that best describes the lake. In some cases, it will be fairly easy to estimate the percentage from observations made during sampling. In other cases, it will be an educated guess, especially if the

**TABLE C.7    SHORELINE CHARACTERISTICS OBSERVED DURING LAKE ASSESSMENT**

| Check Percentage of Shoreline Characteristics | |
|---|---|
| Forest/shrub | Deciduous, coniferous, or mixed forest, including shrub and sapling vegetation |
| Agriculture | Cropland, orchard, feedlot, pastureland, or other horticultural activity |
| Open grass | Meadows, lawns, or other open vegetation |
| Wetland | Forested and nonforested wetlands (submerged terrestrial vegetation) |
| Barren | Nonvegetated areas such as beaches, sandy areas, paved areas, and exposed rock |
| Developed | Immediate shoreline area developed by human activity; this includes lawns, houses, stores, malls, marinas, golf courses, or any other human-built land use |
| Shoreline modifications | Actual shoreline that has been modified by the installation of riprap, revetments, piers, or other human modifications |

*Source:* Herlihy, A.T. 1997. Section 9. Final lake activities. In J.R. Baker, D.V. Peck, and D.W. Sutton (Eds.), *Environmental Monitoring and Assessment Program Surface Waters: Field Operations Manual for Lakes*. Report No. EPA/620/R-97/001. US Environmental Protection Agency, Washington, DC.

water is turbid. After recording the areal percentage of macrophyte coverage, record the density of the plants in the observed macrophyte beds as dense, moderate, or sparse. Finally, provide any qualitative description (genera present [if known], dominant type [floating, emergent, or submergent]) of the macrophyte beds that would be useful for interpreting the trophic status of the lake. All activities described in this section are recorded on the Lake Assessment portion of the Lake-Water-Sampling Record Form.

## C.6  POST-TRIP ACTIVITIES

### C.6.1  Data Forms and Sample Inspection

After the Lake-Water-Sampling Record and Chain-of-Custody Forms are completed, one team member reviews the data forms and sample labels for accuracy, completeness, and legibility. Confirm that the lake ID is correct on the forms, as well as the date of the visit. Verify that all information has been recorded accurately, the recorded information is legible, and any flags are explained in the Comments section. Ensure that written comments are legible and use no "shorthand" or abbreviations. After reviewing the Lake-Water-Sampling Record Form, initial the lower right corner of each page of the form. Ensure that all samples are labeled, all labels are completely filled in, and each label is covered with clear plastic tape.

### C.6.2  Launch Site Cleanup

If a boat or inflatable raft or float tube was used for lake sampling, inspect it for evidence of weeds and other macrophytes. Clean the boat or raft as completely as possible before leaving the launch site to minimize the possibility of transporting aquatic plant fragments or aquatic animals to other lakes where these species may not already occur. Clean up all waste material at the launch site and dispose of it or transport it out of the site.

### C.6.3  Processing Site Documentation Data and Information

A file system and database with reliable backup should be established for storage of site records and files, map images, and photos. Processing of site documentation data and information includes the following steps:

A. Retrieve site coordinates (and any tracks) from the GPS unit using the GPS software. Delete any extra coordinate sets (waypoints) and save the file as a *.gdb file.
B. Retrieve photos from the camera.

C. Enter or revise the site record in the database.
  1. Enter site coordinates obtained in the field.
  2. Enter or revise the site description and travel and access directions.
  3. Add new photos as JPEG images with captions to the site record.
D. Create site maps providing both detailed and broader information for access and orientation. Annotate maps and pictures with text and arrows when it would be helpful. Note that the accuracy of maps varies, and the coordinate-based point on the maps, as well as other information, may be misleading. Add clarifying notes. Save these maps as JPEG images in the site record. Add captions as appropriate.

## C.7  ACKNOWLEDGMENTS

This SOP is based partly on material developed by Baker and Peck (1997), Baker et al. (1997), Herlihy (1997), and Merritt et al. (1997).

## REFERENCES

Baker, J. R., A.T. Herlihy, S.S. Dixit, and R. Stemberger. 1997. Section 7. Water and sediment sampling. In J.R. Baker, D.V. Peck, and D.W. Sutton (Eds.), *Environmental Monitoring and Assessment Program Surface Waters: Field Operations Manual for Lakes*. Report No. EPA/620/R-97/001. US Environmental Protection Agency, Washington, DC.

Baker, J.R. and D.V. Peck. 1997. Section 4. Lake verification and index site location. In J.R. Baker, D.V. Peck, and D.W. Sutton (Eds.), *Environmental Monitoring and Assessment Program Surface Waters: Field Operations Manual for Lakes*. Report No. EPA/620/R-97/001. US Environmental Protection Agency, Washington, DC.

Herlihy, A.T. 1997. Section 9. Final lake activities. In J.R. Baker, D.V. Peck, and D.W. Sutton (Eds.), *Environmental Monitoring and Assessment Program Surface Waters: Field Operations Manual for Lakes*. Report No. EPA/620/R-97/001. US Environmental Protection Agency, Washington, DC.

Merritt, G.D., V.C. Rogers, and D.V. Peck. 1997. Section 3. Base site activities. In J.R. Baker, D.V. Peck, and D.W. Sutton (Eds.), *Environmental Monitoring and Assessment Program Surface Waters: Field Operations Manual for Lakes*. Report No. EPA/620/R-97/001. US Environmental Protection Agency, Washington, DC.

# Appendix D: Data Entry Forms and Labels for Field-Sampling Activities

# STREAM-SAMPLING-SITE DOCUMENTATION FORM

| A) BASIC INFORMATION |
|---|

Stream Name: USGS: _____    Sample Site Name: _____
            Local: _____    Sample Site ID Number: _____

Forest/Wilderness/Park/Other (circle one) Name: _____

Date of Visit: __ __ /__ __ /__ __ __ __    Visit Number: ☐ Initial ☐ Subsequent

Field Team Leader:
Name: _____ Affiliation: _____
Phone: _____ Email: _____

Access:   ☐ Vehicle   ☐ Short Hike (<1 hr)   ☐ Long Hike (>1 hr)   ☐ Overnight Hike

| B) STREAM SITE VERIFICATION |
|---|

Stream verified                  ☐ Yes    ☐ No

Stream site verified by (✓ all that apply)  ☐ GPS   ☐ Local Contact  ☐ Signs   ☐ Vegetation
                               ☐ Roads  ☐ Topo Map     ☐ Photos  ☐ Other

Site Tag Has Been Affixed: ☐ Yes ☐ No ☐ New Tag ☐ Existing Tag

Tag Tree Species _____ Tag Tree Description and Location Relative to Stream
    Sampling Site: _____

| C) GPS COORDINATES |
|---|

|  | Latitude (Decimal Degrees) | Longitude (Decimal Degress) | Resolution (from manual) | Elevation |
|---|---|---|---|---|
| Sample Site: | __ __ . __ __ __ __ __ | -__ __ . __ __ __ __ __ | ____ ☐ m ☐ ft | _____ m |

Do GPS coordinates and elevation correspond to map? ☐ Yes ☐ No Datum:_____
Explain _____

| D) SAMPLING SITE DESCRIPTION AND COMMENTS |
|---|

Travel Directions to Stream Sampling Site and Access Information:




Description of Stream at Sampling Site:




Name (Local/USGS):            Date:        ID:

## E) SAMPLE SITE ASSESSMENT

Dominant Land Use in Vicinity of Sampling Site:
□ Forest/Shrub □ Agriculture □ Open Herbaceous □ Developed □ Wetland

If Known, What are the Dominant Plant Species? _____
_____

If Forest, Dominant Age Class:
  □ 0–10 yrs  10–25 yr □ 25–50 yrs □ >50 yr

Beaver Activity in Vicinity of Sampling Site:
  Beaver Signs: □ None □ Rare □ Common
  Beaver Flow Modifications: □ None □ Minor □ Major

## F) WATERSHED ASSESSMENT (Done through GIS project in office)

Lithology:

Percent of Watershed Above Stream Sampling Site in: Hardwoods ____% Conifers ____%
  Mixed Forest ____% Exposed Bedrock ____% Herbaceous/Shrubs ____% Talus ____%

Watershed Area Above Sample Point: _____ □ $km^2$ □ $mi^2$

Watershed Aspect (degrees) _____°

Average Slope of Watershed: _____%

Stream Order: _____ Data/Method Used to Determine Stream Order _____

## G) PHOTOS

Attach photos with file name, date, and description.

## H) OTHER NOTES AND/OR SITE SKETCH

Completed by (initial) _____

## Stream-Sampling-Site Documentation Form Instructions

**Stream Name:** Enter the name of the stream to be sampled. If applicable, provide both the US Geological Survey (USGS) name (from topo map) and local name that the stream is known by.

**Sample Site Name and ID Number:** Each sample site (location) will have a unique name and ID number assigned to it. These are generated locally for the project. The identification number will appear on all sample bottles used for sampling this stream at this location. It is especially important to have a unique site ID when there is more than one sample location on a stream.

**Forest/Wilderness/Park/Other:** Circle one of these options and write the name of the forest/wilderness/park or other in the space provided.

**Date of Visit:** Enter the date of visiting the stream-sampling site.

**Visit Number:** Check whether this is the first (initial) visit to this site to establish it as part of a survey or monitoring effort or if this site has been visited and documented previously and therefore this is a subsequent visit.

**Field Team Leader:** Enter the name, affiliation, phone number, and e-mail address of the responsible field person.

**Access:** Check the box that best represents the mode of site access.

**Stream Site Verification:** Do the available data match conditions observed on the ground sufficient to verify that the intended sampling site has been located (Yes/No)? Check all methods used to verify that you have located the correct site.

**Tree Tag:** If allowed by applicable regulations, has a metal tree tag been affixed to a prominent tree in proximity to the sampling site? If so, indicate what species, if known. If the species is not known, indicate as such. Was the tag affixed to the tree on a previous trip (existing tag) or newly placed on this trip (new tag)? Describe the tree and its location relative to the sampling site. Include in the description the height above the ground and compass bearing from the tag to the sampling site.

**GPS Coordinates:** Enter the latitude and longitude of the sampling site in decimal degrees (to at least six places after the decimal). Enter from the global positioning system (GPS) manual its approximate resolution and whether that resolution is expressed in meters or feet. Enter the datum used (e.g., NAD83). Use NAD83 if possible; otherwise, you will have to convert it to NAD83 before entering it into the database.

**Sampling Site Description and Comments:** Describe the travel directions and any access issues or difficulties. Describe the sampling site itself.

**Lake Name, Date, ID:** Repeat the lake name (circle whether it is the official USGS name or a local name), date of sampling, and lake ID.

**Sample Site Assessment:** Estimate land use and vegetation within the immediate vicinity of the sample site. What is the dominant land use in the vicinity of the sampling site? What are the dominant plant species (if known to the sampler)? If forest, what is the approximate age class? To what extent have beaver influenced the general vicinity of the sampling site?

**Watershed Assessment:** Most of this information can be obtained through a geographical information system (GIS) project prior to, or after, visiting the sample site. It can be helpful to have this information already filled in prior to the sampling trip. It can help to verify that you have located the targeted sample location or, conversely, to indicate that available mapped information may be in error.

**Photos:** Attach photos of the sampling site looking both upstream and downstream. Attach any additional photos of the stream, sampling-site location, or access route that might be helpful in locating or accessing this site. Label each photo with file name, date photographed, and description of where photo was taken.

**Notes:** Add any additional information that may help to identify, locate, or describe this site. Add a sketch of the site if you think that would be helpful for locating it again in the future.

**Initials:** Place your initials on the bottom of the form.

# STREAM-WATER-SAMPLING RECORD FORM

**A) SITE INFORMATION**

Stream Name: USGS: _____    Sample Site Name: _____
Local: _____    Sample Site ID Number: _____

Date Sampled: _____    Arrival Time (24 hr) __ __ __ __ ☐ Standard ☐ Daylight Savings

Field Team Leader: Name _____    Telephone _____
                    Affiliation _____    Email

_____

**B) SUGGESTED REVISIONS TO STREAM SAMPLING SITE DOCUMENTATION**

☐ GPS Coordinates ☐ Stream Description ☐ Sampling Site Description ☐ Travel Directions
Describe Suggested Revisions:

**C) GENERAL OBSERVATIONS**

| | |
|---|---|
| Air Temperature | ☐ °C ☐ °F Time Measured (24 hr) __ __ __ __ |
| Collection Day Weather up to Time of Sampling | ☐ Clear ☐ Partly cloudy ☐ Overcast<br>☐ Light rain ☐ Occasional rain ☐ Persistent rain<br>☐ Snow or sleet ☐ Hail |
| Weather during Preceding 3 Days | ☐ Generally dry ☐ Occasional rain/snow<br>☐ Generally wet ☐ Very wet |
| Discharge Level | ☐ No flow ☐ Low flow ☐ Normal flow ☐ High flow ☐ Flood |

**D) WATER SAMPLES AND REPLICATES**

Normal Sample:
   Collection Time (24 hour): __ __ __ __
   Sample Identification Code: _____
   Number of Plastic Bottles _____ Number of Syringes _____ Number of Glass Bottles _____

Replicate 1: Collected? ☐ Yes ☐ No
   Collection Time (24 hour): __ __ __ __
   Sample Identification Code: _____
   Number of Plastic Bottles _____ Number of Syringes _____ Number of Glass Bottles _____

Replicate 2: Collected? ☐ Yes ☐ No
   Collection Time (24 hour): __ __ __ __
   Sample Identification Code: _____
   Number of Plastic Bottles _____ Number of Syringes _____ Number of Glass Bottles _____

Approximate Stream Depth in Mid-Channel at Sampling Location: _____ ☐ m ☐ ft ☐ cm ☐ in

Approximate Stream Width at Sampling Location: _____ ☐ m ☐ ft ☐ cm ☐ in

Method of Delivery to Laboratory: ☐ Vehicle ☐ Overnight Shipping ☐ Other Explain: _____

Collection Location: Explain any deviation from targeted sampling location _____

| Name (Local/USGS): | Date: | ID: |
|---|---|---|

## E) *IN SITU* WATER DATA

Time Obtained: __ __ __ __ (24 hr)
Air Temperature: _____ □ °C □ °F   Instrument Used: _____
Water Temperature at Sample Location: _____ □ °C □ °F Instrument Used: _____

Field Instrument Data (Optional):

| Parameter | Value | Equipment (Make/ Model) | Method (EPA/SM/ USGS) Reference |
|---|---|---|---|
| Conductivity | _____uS/cm<br>Corrected to 25 °C?<br>□ Yes □ No | | |
| pH | | | |
| Turbidity | _____NTU | | |
| Dissolved Oxygen | _____mg/l<br>_____%DO | | |
| Other: | _____ (____) | | |

## F) PHOTOS

Attach photos with file name, date, and description.

## G) NOTES

| Name (Local/USGS): | Date: | ID: |
|---|---|---|

## H) STAGE AND DISCHARGE DATA

1. General Information

Time Obtained (24 hr): ___ ___ ___ ___   □ Not Obtained in Field
Method to Determining Stage and/or Discharge (Check [✔] all that apply): □ None
   □ Cross section of depth measurements
   □ Velocity-area procedure (number of sets of measurements taken _____)
   □ Salt dilution method
   □ Relative stage comparision with nearby fixed gage
   □ Stage measurement with pressure transducer
   □ Stage measurement with staff gage
   □ Timed filling procedure (number of spillways measured _____)

2. Sage Measurement Only

Stage Relative to Fixed Gage _____ □ m □ ft □ cm □ in
Location of Fixed Gage Measurement Relative to Stream Sampling Site _____
Stage Measurement with Staff Gage _____ □ m □ ft □ cm □ in
Location of Fixed Gage Measurement Relative to Stream Sampling Site _____
Existing Rating Curve: □ yes □ no If yes, referenced to what? _____

Gage Height Referenced to Rating Curve: _____ □ m □ ft □ cm □ in

3. Discharge Measurement by Velocity-Area Procedure

Velocity-Area Procedure: □ yes □ no   Depth: □ m □ ft □ cm □ in   Velocity: □ m/s □ ft/s
Stream Width at Measurement Location _____ □ m □ ft □ cm □ in

| Interval | 1 | 2 | 3 | 4 | 5 | 6 | 7 | 8 | 9 | 10 |
|----------|---|---|---|---|---|---|---|---|---|----|
| Depth | __ | __ | __ | __ | __ | __ | __ | __ | __ | __ |
| Velocity | __ | __ | __ | __ | __ | __ | __ | __ | __ | __ |

| Interval | 11 | 12 | 13 | 14 | 15 | 16 | 17 | 18 | 19 | 20 |
|----------|----|----|----|----|----|----|----|----|----|----|
| Depth | __ | __ | __ | __ | __ | __ | __ | __ | __ | __ |
| Velocity | __ | __ | __ | __ | __ | __ | __ | __ | __ | __ |

4. Discharge Measurement by Timed Filling Procedure

Timed Filling Procedure: □ yes □ no   Time: □ min □ sec   Volume: □ L □ gal

Time to Fill Measured Volume

| Spillway Number | Trial 1 | | Trial 2 | | Trial 3 | | Trial 4 | | Trial 5 | |
|-----------------|------|--------|------|--------|------|--------|------|--------|------|--------|
| | Time | Volume | Time | Volume | Time | Volume | Time | Volume | Time | Volume |
| 1 | __ | __ | __ | __ | __ | __ | __ | __ | __ | __ |
| 2 | __ | __ | __ | __ | __ | __ | __ | __ | __ | __ |
| 3 | __ | __ | __ | __ | __ | __ | __ | __ | __ | __ |

Initials _____

# Stream-Water-Sampling Record Form Instructions

**Stream Name:** Enter the name of the stream to be sampled. Provide both the USGS name (from topo map) and local name that the stream is known by.

**Sample Site Name and ID Number:** Each sample site (location) will have a unique name and ID number assigned to it. These are generated locally for the project. The identification number will appear on all sample bottles used for sampling this stream at this location. It is especially important to have a unique site ID when more than one sample location is located on a stream.

**Date Sampled:** Enter the date of visiting the sampling site.

**Arrival Time:** Indicate time of arrival at sampling location. Use 24-h (military time) format. Indicate whether standard local time or daylight savings local time.

**Field Team Leader:** Enter the name, affiliation, phone number, and e-mail address of the responsible field person.

**Suggested Revisions to Site Documentation:** Examine information given on the Stream-Sampling-Site Documentation Form and indicate any suggested revisions. Place a check mark (✓) in any box requiring revision and explain the suggested revision in the space provided.

**Air Temperature:** Enter air temperature to the nearest degree. Record whether expressed in degrees centigrade or degrees Fahrenheit.

**Weather:** Check the boxes that best describe the collection day weather up to the time of sampling and the average weather over the previous 3 days (if known).

**General Discharge Level:** Indicate the general level of discharge in the stream at the time of sampling.

**Sample ID:** Enter the unique identification code assigned to the sample (ideally, prepared bar codes). The sample ID represents a sample of water (bottle[s] or syringe[s]) intended to represent conditions at a particular location, on a particular day, at a particular time. Note that multiple containers (bottle[s] and syringe[s]) obtained within one time window represent the **same** sample and receive the **same** ID code. Replicated samples receive different ID codes. The ID code may be prepared as a computer-generated bar code. Multiple stick-on copies of the bar code can be prepared prior to field sampling and subsequently be affixed to the Stream-Water-Sampling Record Form, the Chain-of-Custody Form, and each container (bottle or syringe) for the sample.

**Collection Time:** Indicate time of routine sample collection, using a 24-h clock (thus, 4 pm is 1600). *Note that the time recorded on the bottle[s] and syringe[s] for the replicates should differ from the time recorded for* *the normal sample. **This is important.*** The recommended protocol is to separate the sampling times for normal and replicate samples by 1 min.

**Samples:** Indicate the number and type of sample aliquots collected for the normal (routine) sample and any replicates that may have been collected. How was the sample shipped or transported to the lab? Explain any deviation from the intended sampling location.

***In Situ* Water Data:** If any *in situ* water data were collected, provide the measured values. At what time were the *in situ* measurements taken? Indicate whether temperatures are expressed in Celsius or Fahrenheit. Express dissolved oxygen (DO) in units of milligrams per liter (mg/L) and, if possible, percentage DO. Correct specific conductance to 25°C. Indicate the instruments and methods used.

**Photos:** Attach photos of sampling site looking both upstream and downstream. Label each photo with file name, date photographed, and description of where photo was taken. Attach and describe other photos, as appropriate.

**Notes:** Add any additional information that may help to identify, locate, or describe this site. If chlorophyll *a* sample aliquots were filtered in the field, prior to analysis of the filter in the laboratory, it is important to include in the Notes section of the form the volume of water filtered.

**Stage and Discharge—General Information:** Indicate what methods were used to obtain an estimate of stream stage or discharge. Fill in the information for either 2 (Stage Measurement Only) *or* 3 (Velocity-Area Procedure) *or* 4 (Timed Filling Procedure).

**Stage Measurement:** If stage measurements (estimates) were made in the field, record the measured value and indicate the unit of measure. Describe the location of measurement. Indicate if a rating curve has been developed with which to estimate discharge from stage measurements at this location and indicate what the stage is referenced to (i.e., fixed staff gauge, permanent landscape feature).

**Discharge: Velocity-Area Procedure:** If the velocity-area procedure was used to measure discharge, check "yes" and indicate the units of measurement for water depth and velocity. Record the approximate width of the stream at the sampling location. Record the water depth and velocity in each of up to 20 evenly spaced intervals of the stream cross section.

**Discharge—Timed Filling Procedure:** If the timed filling method was used to measure discharge, check "yes" and indicate the units of measurement for time and water volume. Record the time and volume measurements for five separate trials at each of up to three spillway locations.

**Initials:** Place your initials on the bottom of the form.

# LAKE-SAMPLING-SITE DOCUMENTATION FORM

| A) BASIC INFORMATION | | |
|---|---|---|
| USGS Name: | Local Name: | ID: |

Forest/Wilderness/Park/Other (circle one) Name: _____

| Date of Visit: _ _ / _ _ / _ _ _ _ | Elevation: _____ □ m □ ft | Lake Area: ____ □ ha □ ac |
|---|---|---|

Field Team Leader:
Name: _____  Affiliation: _____
Phone: _____  Email: _____

Access: □ Vehicle □ Short Hike (<1 hr) □ Long Hike (>1 hr) □ Overnight Hike

Lake Determined to Be Non-Target? □ Yes □ No
If Yes, Reason: □ Small Area □ Shallow □ Weedy □ Other: _____

## B) LAKE VERIFICATION AND COORDINATES

Lake verified by:  □ GPS  □ Local Contact  □ Signs  □ Vegetation
(✓ all that apply)  □ Trails/Roads  □ Topo Map  □ Photos  □ Other

Sample Site:  Lat (DD) _ _ . _ _ _ _ _ _  Long (DD)- _ _ _ . _ _ _ _ _ _  Datum: _____

## C) LAKE ASSESSMENT

1. General Lake Hydrologic Information

Hydrologic Lake Type: □ Reservoir □ Drainage (Outlets Present) □ Seepage (No Outlets Observed)

| # Inlets _____ # Outlets _____ | Outlet Dams: □ None □ Artificial □ Augmented □ Natural |
|---|---|

Reference Point Lake Level Changes:  □ Zero □ Change = +/− _____ □ m □ ft
Lake Level Location _____  Reference Point Description _____

Discharge Level Estimate: □ No flow □ Low flow □ Normal flow □ High flow □ Flood

2. Shoreline Characteristics (% of shoreline)

| | | | | |
|---|---|---|---|---|
| Forest/Shrub | □ Rare (<5%) | □ Sparse (5 to 25%) | □ Moderate (25 to 75%) | □ Extensive (>75%) |
| Open Herbaceous | □ Rare (<5%) | □ Sparse (5 to 25%) | □ Moderate (25 to 75%) | □ Extensive (>75%) |
| Wetland | □ Rare (<5%) | □ Sparse (5 to 25%) | □ Moderate (25 to 75%) | □ Extensive (>75%) |
| Barren (Beach or Rock) | □ Rare (<5%) | □ Sparse (5 to 25%) | □ Moderate (25 to 75%) | □ Extensive (>75%) |
| Agriculture | □ Rare (<5%) | □ Sparse (5 to 25%) | □ Moderate (25 to 75%) | □ Extensive (>75%) |
| Developed | □ Rare (<5%) | □ Sparse (5 to 25%) | □ Moderate (25 to 75%) | □ Extensive (>75%) |
| Shoreline Mods. (Docks, Riprap) | □ Rare (<5%) | □ Sparse (5 to 25%) | □ Moderate (25 to 75%) | □ Extensive (>75%) |

| Name (Local/USGS): | Date: | ID: |
|---|---|---|

| 3. Macrophyte Survey    Macrophyte Density | □ Absent | □ Sparse | □ Moderate | □ Dense |
|---|---|---|---|---|
| Emergent/Floating Coverage (% Lake Area) | □ 0 to 25% | □ 25 to 50% | □ 50 to 75% | □ > 75% |
| Submergent Coverage (% Lake Area) | □ 0 to 25% | □ 25 to 50% | □ 50 to 75% | □ > 75% |

Name(s) of one to three of the most prevalent macrophyte species (if known):

Macrophyte description:

4. Lake Trophic State Estimation (based on amount of biomass in lake)
□ Oligotrophic (little/none) □ Mesotrophic (intermediate) □ Eutrophic (large amt)
□ Hypereutrophic (extreme)

## D) WATERSHED ASSESSMENT (Through GIS project in office)

Lithology:

Watershed coverage: ____% Glacier/perm. snow ____% Exposed Bedrock ____% Talus
____% Hardwoods ____% Conifers ____% Mixed Forest ____% Herbaceous/Shrubs

Watershed Area Above Lake: ____ □ km² □ mi² Primary Aspect (degrees) _____ °

Average Slope of Watershed: _____%

## E) SAMPLING SITE LOCATION AND DIRECTIONS

Site Description: □ Index Site □ Lake Outlet □ Shoreline □ Lake Inlet □ Other _____
Describe launch site and location of sample collection:

Directions to lake:

## F) OTHER NOTES

| Name (Local/USGS): | Date: | ID: |
|---|---|---|

## G) LAKE SKETCH

Provide "North" arrow    Mark site:    L = Launch           I = Index Site
                                       S = Shoreline Site    O = Outlet Site

## G) PHOTOS

Attach photos with file name, date, and description.

Completed by (initial) _____

## Lake-Sampling-Site Documentation Form Instructions

**Lake Names and ID Number:** Enter the official USGS name of the lake to be sampled, the local name if applicable, and the identification number assigned to it. This identification number will appear on all sample bottles used for sampling this lake at this location.

**Forest/Wilderness/Park/Other:** Circle one of these options and write the name of the forest/wilderness/park or other in the space provided.

**Date of Visit:** Enter the date of visiting the lake-sampling site.

**Visit Number:** Check whether this is the first (initial) visit to this site to establish it as part of a survey or monitoring effort or if this site has been visited and documented previously and therefore this is a subsequent visit.

**Elevation:** From GIS or mapped data available in the office, what is the elevation of this lake? Is it expressed in meters or feet? Having these data available in advance of the site visit will help verify that you have found the correct lake.

**Lake Area:** From GIS or mapped data available in the office, what is the approximate area of this lake? Is it expressed in hectares or acres? Having these data available in advance of the site visit will help verify that you have found the correct lake.

**Field Team Leader:** Enter the name, affiliation, phone number, and e-mail of the team leader.

**Access:** Check the box that best represents the mode of site access.

**Nontarget Determination:** Check (✓) if the lake was determined to be nontarget and therefore will not be sampled. If applicable, indicate the reason for designating the lake as nontarget (i.e., too small, too shallow, too weedy, or some other reason).

**Lake Verification:** Check (✓) all means used to verify the lake identity.

**Coordinates:** Enter the latitude and longitude in decimal degrees (to at least six places after the decimal) of the sampling site. Enter the datum used for the coordinates. Use NAD83 if possible; otherwise, you will have to convert it to NAD83 before entering it into the database.

**Lake Assessment:** Indicate the hydrologic type of the lake based on the presence of an artificial human-made dam (reservoir) or as a drainage lake (outlet stream present; may or may not be flowing at time of visit) or seepage lake (no outlet stream present, regardless of whether it is flowing at the time of visit).

**# Inlets/# Outlets:** Indicate the number of inlets and outlets to the lake.

**Dam:** Is there a dam present? If so, is it an artificial structure that creates, or largely creates, the subject lake? Is it a smaller feature that serves to raise the stage of a previously existing lake (augmented)? Is it a natural dam, such as from a rockslide or lava flow?

**Reference Point Lake Level Changes:** Record change in the lake level relative to the fixed gauge or point (if present). Describe where the measurement was taken and the reference point.

**Discharge Level Estimate:** Indicate the general level of discharge in the lake outlet stream (if present) at the time of sampling.

**Shoreline:** Indicate the relative percentages of vegetation types around the shoreline to a distance of about 20 m from the lake.

**Lake Name/Date/ID:** Repeat lake name (circle whether it is the official USGS name or a local name), date of sampling, and lake ID.

**Macrophytes:** From a quick visual survey of the lake, what is the average density of macrophytes (aquatic plants large enough to be seen without magnification)? Estimate the percentage coverage of emergent and submergent macrophytes. Identify (if known) the one to three most prevalent macrophyte species. Describe the general macrophyte community.

**Lake Trophic State:** Based on the amount of biomass in the lake, indicate your estimation of the lake's trophic state.

**Watershed Assessment:** Most of this information can be obtained through a GIS project prior to, or after, visiting the sample site. It can be helpful to have this information already filled in prior to the sampling trip. It can help to verify that you have located the targeted sample location or, conversely, to indicate that available mapped information may be in error. *Lithology*: If you have GIS coverage of bedrock geology/lithology, indicate the primary lithology type in the watershed above the lake. *Cover*: Percentages in cover types indicated; The totals must equal 100%. If you know actual vegetation cover types, use them instead. *Watershed area*: The area of the watershed feeding the lake in square kilometers. *Primary aspect:* The primary aspect of the watershed as indicated in degrees with north 0°, east 90°, south 180°, and west 270°. *Average slope*: Average slope of the catchment in percent.

**Site Location:** Check whether the sample was collected from the index site, lake outlet, shoreline site, or lake inlet. Describe the sampling site and (if applicable) the launch site.

**Directions:** Describe how to get to the lake, including any access issues or difficulties.

**Notes:** Add any additional information that may help to identify, locate, or describe this site.

**Lake Name/Date/ID:** Repeat lake name (circle whether it is the official USGS name or a local name), date of sampling, and lake ID.

**Lake Sketch:** Provide a sketch of the lake, showing the major inlets and outlets. Include a north arrow (labeled *N*) and arrows to indicate the flow direction of the tributary streams. Mark the sampling site (index [I], shoreline [S], outlet [O], or inlet [IL]). If index, also indicate raft/float tube launch location (L).

**Photos:** Attach photos of the sampling site looking both upgradient and downgradient. Label each photo with file name, date photographed, and description of where photo was taken. Attach and describe any other photographs taken at the sampling site or along the access route.

**Initials:** Place your initials on the bottom of the form.

# LAKE-WATER-SAMPLING RECORD FORM

## A) SITE INFORMATION

USGS Name: _____    Local Name: _____    ID: _____
Data Sampled: _____    Arrival Time (24 hr): _ _ _ _    □ Standard
                                                                  □ Daylight Savings

Field Team Leader:  Name _____    Telephone _____
                    Affiliation _____    Email _____

## B) SAMPLING LOCATION AND PROTOCOL DEVIATIONS

Collection Location: □ Index Site □ Shoreline □ Outlet □ Other _____

Sample Depth _____ □ m □ ft
Zone: □ Surface □ Epilimnion □ Hypolimnion □ Deep □ Other _____

Protocol or Target Site Deviation? □ No □ Yes Explain:

## C) GENERAL OBSERVATIONS

Air Temperature        □ °C □ °F

| Collection Day Weather | □ Clear | □ Partly Cloudy | □ Overcast |
| | □ Light Rain | □ Occasional Rain | □ Persistent Rain |
| | □ Snow or Sleet | □ Hail | |

| Weather during Preceding 3 Days | □ Generally Dry | □ Occasional Showers | □ Now Known |
| | □ Generally Wet | □ Very Wet | |

## D) WATER SAMPLES

Normal:  # Plastic Bottles _____ # Glass Bottles _____ # Syringes _____

Sample Identification Code: _____ Collection Time (24 hour): _ _ _ _

Replicate 1 Type: Duplicate/Field Split (circle one) # Plastic Bottles ____ # Glass ____ # Syringes ____

Sample Identification Code: _____ Collection Time (24 hour): _ _ _ _

Replicate 2 Type: Triplicate/Field Split (circle one) # Plastic Bottles ____ # Glass ____ # Syringes ____

Sample Identification Code: _____ Collection Time (24 hour): _ _ _ _

| Name (Local/USGS): _____ | Date: | ID: |

## F) FIELD WATER DATA

Time Obtained: __ __ __ __ (24 hr)          Air Temperature: _____ ☐ °C ☐ °F

Secchi Depth: Depth Disappeared _____ ☐ m ☐ ft    ☐ Clear to Bottom? ☐ Yes ☐ No
              Depth Appeared _____ ☐ m ☐ ft

Water Temp: _____ ☐ °C ☐ °F Equip (Make/Model): _____

Field Measurements

| Parameter | Value | Method | Equipment (Make/Model) |
|---|---|---|---|
| Conductivity (uS/cm) (Corrected to 25°C? ☐ Yes ☐ No) | —— ———— | EPA/SM/USG | _____ |
| pH | ____ _____ | EPA/SM/USG | _____ |
| Turbidity (NTU) | ____ _____ | EPA/SM/USG | _____ |
| DO (units: ☐ % ☐ mg/L) | ____ _____ | EPA/SM/USG | _____ |
| Other: _____ units: ___ | ____ _____ | EPA/SM/USG | _____ |

Depth Profile: Method _____ Instrument Make/Model: _____
Index Location? ☐ Yes ☐ No If no, describe: _____
Depth: ☐ m ☐ ft Conductivity corrected to 25°C: ☐ Yes ☐ No Temperature: ☐ °C ☐ °F
DO units: ☐ % ☐ mg/L Turb = NTU

| Depth | Temp | DO | pH | Cond | Turb | Depth | Temp | DO | pH | Cond | Turb | Depth | Temp | DO | pH | Cond | Turb |
|---|---|---|---|---|---|---|---|---|---|---|---|---|---|---|---|---|---|
| | | | | | | | | | | | | | | | | | |

## H) NOTES

Completed by (initial)

# Lake-Water-Sampling Record Form Instructions

**Lake Names and ID Number:** Enter the official USGS name of the lake to be sampled, the local name if applicable, and the identification number assigned to it. This identification number will appear on all sample bottles used for sampling this lake at this location.

**Date Sampled:** Enter the date of visiting the lake-sampling site.

**Arrival Time:** Indicate time of arrival at sampling location. Use 24-h clock format.

**Field Team Leader:** Enter the name, affiliation, phone number, and e-mail of the team leader.

**Collection Location:** Check the box for the location of this sample and its replicates. If you are collecting the sample from the lake outlet (or inlet), use the Stream-Water-Sampling Protocols for instructions regarding how to collect the sample.

**Sample Depth:** Write the depth at which the sample was collected. Indicate the limnetic zone of the collection. If you do not know, check either Surface or Deep.

**Protocol or Target Site Deviation:** If there were any deviations from the protocols, standard operating procedures (SOPs), or planned site of collection, check Yes and explain these deviations.

**Air Temperature:** Enter air temperature to the nearest degree and the unit of measure.

**Weather:** Check the boxes that best describe the collection day weather and the average weather over the previous 3 days.

**Sample ID:** Enter the unique identification code assigned to the sample. The sample ID represents a sample of water (bottle[s] or syringe[s]) intended to represent conditions at a particular location, on a particular day, at a particular time. Note that multiple containers (bottles and syringes) obtained within one time window represent the **same** sample and receive the **same** ID code. Replicated samples and their syringes receive different ID codes. The ID code may be received from the lab or prepared as a computer-generated bar code label prior to field sampling and then affixed to the Lake-Water-Sampling Record Form, the Chain-of-Custody Form, and each container (bottle or syringe) for the sample.

**Collection Time:** Indicate time of normal sample collection, using a 24-h clock. Note that the time recorded on the bottle(s) and syringe(s) for the second sample of a duplicate pair or for the second and third samples of a triplicate set should differ from the time recorded for the routine sample. The recommended protocol is to separate the sampling times for replicate samples by 1 min.

**Lake Name/Date/ID:** Repeat lake name (circle whether it is the official USGS name or a local name), date of sampling, and lake ID.

**Field Water Data:** *Time Obtained*: Time the field water measurements were taken. *Air Temperature:* Record air temperature and whether it is in degrees Fahrenheit or degrees centigrade. *Secchi Depth:* Record both the depth of the Secchi disk disappearance and reappearance and unit of measurement. If it was clear all the way to the bottom of the lake, check the box. *Water Temperature:* Record water temperature and whether it is in degrees Fahrenheit or degrees centigrade as well as what equipment (with make and model) was used. *Field Measurements:* For each field measurement taken, enter the value, method used (circle Environmental Protection Agency [EPA], SM [for standard method], or USGS), and the equipment make and model used for the measurement. For conductivity, check whether the value was corrected to 25°C. If you measure DO, check the box for the unit of measure. Indicate the parameter in Other (e.g., $NO_3$) for any other parameters measured.

**Depth Profile:** Enter (if known) the EPA/SM/USGS method used for the measurements, as well as the make and model of the equipment used for the profile. Indicate if the depth profile was measured at the index site or at some other location (describe). What is the total depth at the location of the profile? Check the appropriate box for units of measure for temperature and DO. Indicate whether conductivity measurements are corrected to 25°C. At each measured depth, provide the depth and results of each measurement taken at that depth.

**Notes:** Add any additional information about the sample collection and measurements that might help in understanding results. If chlorophyll *a* sample aliquots were filtered in the field, prior to analysis in the laboratory of the filter, it is important to include in the Notes section of the form the volume of water filtered.

**Photos:** Attach photos of sampling site looking both upgradient and downgradient. Label each photo with file name, date photographed, and description of where photo was taken. Attach and describe other photos, as appropriate.

**Initials:** Place your initials on the bottom of the form.

# CHAIN-OF-CUSTODY FORM

| Date Sampled | Time Sampled (24-hour) □ Standard □ Daylight Savings | Sample ID (Barcode) | Sample Location Lake/Stream Name or Latitude/Longitude | Sample Type (Normal, Rep 1, Rep 2, Blank, Split) | Filtered (Y/N) Where? (Field or Field Lab) | Preserved (Y/N/Type) | Analyses Requested | Lab ID Assigned |
|---|---|---|---|---|---|---|---|---|
| _/_/_ | _____ | | | | | | | |
| _/_/_ | _____ | | | | | | | |
| _/_/_ | _____ | | | | | | | |
| _/_/_ | _____ | | | | | | | |
| _/_/_ | _____ | | | | | | | |
| _/_/_ | _____ | | | | | | | |

Forest/Wilderness/Park/Other (Circle One)

Name:

Contact Individual and Affiliation:

Address

Phone Number

Shipped to (Lab Name and Address):

Lab Phone #
Lab Contact
Lab Email

Shipped by: UPS/FedEx/USPS/Other
Shipping #

Page ___ of ___

Comments:

Received/Relinquished by:

| Print Name | Signature | Date & Time Relinquished | Date & Time Received |
|---|---|---|---|

Received at Laboratory by:

| Print Name | Signature | Date Received | Time Received |
|---|---|---|---|

## Chain-of-Custody Form Instructions

**Forest/Wilderness/Park/Other (Circle one):** Circle one of these options and write the name of the forest/wilderness/park or other in the space provided. Provide the name and affiliation of the project contact individual.

**Address and Phone Number:** Provide address and phone number of the forest, wilderness, park, or other. Please include the city, state, and zip code.

**Shipped to (Lab Name):** Name, address, and e-mail of the laboratory where the water samples and original Chain-of-Custody Form will be sent.

**Lab Phone #:** Phone number of the laboratory where the water samples and original Chain-of-Custody Form will be sent.

**Lab Contact and E-mail:** Contact person and e-mail address in the analytical laboratory.

**Shipped by UPS/Fed Ex/USPS/Other:** Identify the carrier you used. **Remember to consider the arrival date of the shipped samples because on weekends and government holidays, there may not be anyone to receive samples in the laboratory.** In general, you should try to ship samples on Monday, Tuesday, or Wednesday.

**Shipping #:** Tracking number assigned to the shipment by the carrier.

**Page ___ of ___:** Page number(s) of Chain-of-Custody Form sent.

**Date Sampled:** Date sample was taken (mm/dd/yyyy).

**Time Sampled:** Time sample was taken (24 h __ __ __ __). Indicate whether standard or daylight savings time.

**Sample ID:** Unique identification number assigned to the sample in the field for tracking purposes (i.e., bar code or unique identifier used to identify the sample for connectivity of field analysis with the laboratory analysis). Ideally, multiple stick-on copies of the sample ID label are prepared as a computer-generated bar code that can be affixed in the field to multiple forms and sample containers.

**Location:** Location description of sample site, including lake or stream name, latitude and longitude in decimal degrees.

**Sample Type (normal, blank, replicate, etc.):** Type of sample collected, such as normal water sample, field blank, replicate, and so on. Replicate (R1, R2) samples are collected at the same location as the normal water sample (N) but at slightly different times (typically 1 min apart). Replicates are usually collected for quality assurance purposes or as backup samples should the normal sample be lost or damaged. A field *blank* is a prepared sample of deionized water that is carried into the field and then shipped to the laboratory with the samples; a field *split* is the second bottle when a normal sample has been split in the field into two bottles. The first bottle is labeled as the normal sample; the second is labeled as the field split (S).

**Filtered Where:** Was this sample filtered in either the field or field laboratory? If so, where?

**Preserved (Y/N):** Was this sample preserved in the field and, if so, with what kind of preservative (e.g., $H_2SO_4$)?

**Analyses Requested:** Instructions for the laboratory requesting what type of analyses are to be performed (e.g., acid-neutralizing capacity [ANC], pH, conductivity, major cations, anions, etc.). You may write "same as usual" if you have an agreement with the laboratory for routine analyses.

**Lab ID:** A unique identifier created by the laboratory to track the samples (if they are not using the sample ID).

**Comments:** Any extra remarks or instructions are placed in this space.

**Received/Relinquished by:**

    **Print Name:** Printed name of sampler relinquishing the samples to another person for shipment to the laboratory or directly to the laboratory.

    **Signature:** Sampler's signature relinquishing the samples to another person for shipment to the laboratory or directly to the laboratory.

    **Date & Time Relinquished:** Date and time relinquished by the sampler or by person shipping samples to the laboratory.

    **Date & Time Received:** Date and time samples were received from the sampler.

**Received at Laboratory by:**

    **Print Name:** Printed name of laboratory personnel receiving the samples.

    **Signature:** Signature of the laboratory personnel receiving the samples.

**Date:** Date the samples were received by the laboratory.

**Time:** Time the samples were received by the laboratory.

**Sender: It is extremely important to send this form and accompanying Lake- or Stream-Water-Sampling Record Form to the laboratory so that proper connections can be made between field and laboratory information and so that relevant data may be entered into the database.**

## Labels for Water, Zooplankton, and Stream Benthic Macroinvertebrate Samples

### Water Sample

Lake or Stream Name: __
Site ID Number: ___
Sample Date: _
Sample Time (24 h): ___ ___ ___ ___
Collected by: ____
Sample Type (check one):
    ☐ Normal (N)    ☐ Rep 1 (R1)    ☐ Rep 2 (R2)
    ☐ Field Blank    ☐ Field Split

Container Type (check one):
    ☐ Plastic bottle    ☐ Glass bottle    ☐ Syringe

Sample ID/Bar Code:

```
1  0  T  S  1  1  0  6
```

### Zooplankton Sample

Lake Name: _
Site ID Number: _
Sample Date: ____
Sample Time (24 h): ___ ___ ___ ___
Collected by: ____
Sample Depth    ☐ m    ☐ ft: __
Net Mesh Size:    ☐ 80 μm    ☐ 243 μm    ☐ Other
Specify _
Sample ID/Bar Code:

```
1  0  T  S  1  1  0  6
```

### Stream Benthic Macroinvertebrate Sample

Stream Name: ___
Site ID Number: _
Sample Date: _
Sample Time (24 h): ___ ___ ___ ___

Collected by: __
Number of Kick Net Samples Collected: ___
Sample ID/Bar Code:

1  0  T  s  1  1  0  6

## Label Instructions

### Water Sample

**Lake or Stream Name:** Enter the name of the lake or stream sampled. Provide both the USGS name (from topo map) and local name that the lake or stream is known by.

**Site ID Number:** Each sample site (location) will have a unique name and ID number assigned to it. These are generated locally for the project. The identification number will appear on all sample bottles used for sampling this lake or stream at this location. It is especially important to have a unique site ID when more than one sample site is located on a given lake or stream.

**Sample Date:** Enter the date of visiting the sampling site.

**Sample Time (24 h):** Indicate time of arrival at sampling location. Use 24-h (military time) format. Indicate whether it is recorded in standard local time or daylight savings local time.

**Collected by:** Enter the name and affiliation of the responsible field person.

**Sample Type:** Check the box indicating whether this is a normal sample, a replicate sample (Replicate 1 or Replicate 2), or a field blank or field split sample.

**Container Type:** Check whether the water has been collected into a plastic or glass bottle or into a syringe.

**Sample ID/Bar Code:** Enter the unique identification code assigned to the sample (ideally prepared bar code). The sample ID represents a sample of water (bottle[s] or syringe[s]) intended to represent conditions at a particular location, on a particular day, at a particular time. Note that multiple containers (bottle[s] and syringe[s]) obtained within one time window represent the **same** sample and receive the **same** ID code. Replicated samples receive different ID codes. The ID code may be prepared as a computer-generated bar code. Multiple stick-on copies of the bar code can be prepared prior to field sampling and subsequently be affixed to the Stream-Water-Sampling Record Form, the Chain-of-Custody Form, and each container (bottle or syringe) for the sample.

## Zooplankton Sample

**Lake Name:** Enter the name of the lake sampled. Provide both the USGS name (from topo map) and local name that the lake is known by.

**Site ID Number:** Each sample site (location) will have a unique name and ID number assigned to it. These are generated locally for the project. The identification number will appear on all sample bottles used for sampling this lake at this location. It is especially important to have a unique site ID when more than one sample site is located on a given lake.

**Sample Date:** Enter the date of visiting the sampling site.

**Sample Time (24 h):** Indicate time of arrival at sampling location. Use 24-h (military time) format. Indicate whether it is recorded in standard local time or daylight savings local time.

**Collected by:** Enter the name and affiliation of the responsible field person.

**Sample Depth:** Record the water depth at which the sample was collected. Check whether the depth is recorded in meters or feet.

**Net Mesh Size:** Check whether the mesh size of the net used to collect the sample was 80 μM, 243 μM, or some other size. If the mesh was of a size other than 80 or 243 μM, specify the mesh size used.

**Sample ID/Bar Code:** Enter the unique identification code assigned to the sample (ideally prepared bar code). The sample ID represents a sample of water (bottle[s] or syringe[s]) intended to represent conditions at a particular location, on a particular day, at a particular time. Note that multiple containers (bottle[s] and syringe[s]) obtained within one time window represent the **same** sample and receive the **same** ID code. Replicated samples (which typically differ by about 1 min in the time of sampling) receive different ID codes. The ID code may be prepared as a computer-generated bar code. Multiple stick-on copies of the bar code can be prepared prior to field sampling and subsequently be affixed to the Lake-Water-Sampling Record Form, the Chain-of-Custody Form, and each container (bottle or syringe) for the sample.

## Stream Benthic Macroinvertebrate Sample

**Stream Name:** Enter the name of the stream sampled. Provide both the USGS name (from topo map) and local name that the stream is known by.

**Site ID Number:** Each sample site (location) will have a unique name and ID number assigned to it. These are generated locally for the project. The identification number will appear on all sample bottles used for sampling this stream at this location. It is especially important to have a unique site ID when more than one sample site is located on a given stream.

**Sample Date:** Enter the date of visiting the sampling site.

**Sample Time (24 h):** Indicate time of arrival at sampling location. Use 24-h (military time) format. Indicate whether it is recorded in standard local time or daylight savings local time.

**Collected by:** Enter the name and affiliation of the responsible field person.

**Number of Kick Net Samples Collected:** Record the number of discrete kick net samples that were collected and pooled to form this one sample submitted to the laboratory.

**Sample ID/Bar Code:** Enter the unique identification code assigned to the sample (ideally prepared bar code). The sample ID represents a sample of water (bottle[s] or syringe[s]) intended to represent conditions at a particular location, on a particular day, at a particular time. Note that multiple containers (bottle[s] and syringe[s]) obtained within one time window represent the **same** sample and receive the **same** ID code. Replicated samples receive different ID codes. The ID code may be prepared as a computer-generated bar code. Multiple stick-on copies of the bar code can be prepared prior to field sampling and subsequently be affixed to the Stream-Water-Sampling Record Form, the Chain-of-Custody Form, and each container (bottle or syringe) for the sample.

# STREAM BENTHIC MACROINVERTEBRATE SAMPLING DATA FORM*

Site ID: _____    Stream Name: _____    Sample Site Name: _____    Sample ID: _____

Location:    Latitude ___ . ___    Longitude: ___ . ___    Datum: _____

Sample Date: _____    Sample Time (24-hr): ___ : ___    □ Standard □ Daylight Savings

Field Team Leader:    Name _____    Telephone _____
                      Affiliation _____    Email _____

Water Sample for Chemical Analysis Collected on the Same Day as Macroinvertebrate Sample? □ Yes □ No
Has a Stream Sampling Site Documentation Form Been Completed for This Site? □ Yes □ No If not, complete one now.
Number of Jars: □ 1 □ 2 □ 3 □ Other
Sample Replicated: □ Yes □ No  Replicate Sample ID: _____
Number of Kick Net Samples Collected and Combined: _____

Substrate and Channel Types (check all boxes that apply):

| Transect | A | | B | | C | | D | | E | | F | | G | | H | | I | | J | | K | |
|---|---|---|---|---|---|---|---|---|---|---|---|---|---|---|---|---|---|---|---|---|---|
| Substrate / Channel | Sub | Chan | Sub | Chan | Sub | Chan | Sub | Chan | Sub | Chan | Sub | Chan | Sub | Chan | Sub | Chan | Sub | Chan | Sub | Chan | Sub | Chan |
| Fine/Sand | Pool | □ | □ | □ | □ | □ | □ | □ | □ | □ | □ | □ | □ | □ | □ | □ | □ | □ | □ | □ | □ | □ |
| Gravel | Glide | □ | □ | □ | □ | □ | □ | □ | □ | □ | □ | □ | □ | □ | □ | □ | □ | □ | □ | □ | □ |
| Coarse | Riffle | □ | □ | □ | □ | □ | □ | □ | □ | □ | □ | □ | □ | □ | □ | □ | □ | □ | □ | □ | □ |
| Other: Note in Comments | Rapid | □ | □ | □ | □ | □ | □ | □ | □ | □ | □ | □ | □ | □ | □ | □ | □ | □ | □ | □ | □ |

Field Comments:

Channel Types:
Pool–still water; low velocity
Glide–water moving slowly w/smooth surface
Riffle–Small ripples and waves on surface
Rapid–Turbulent water movement w/continuous rushing sound

Substrate Size Classes:
Fine/Sand–ladybug or smaller (<2 mm)
Gravel–ladybug to tennis ball (2–64 mm)
Coarse–tennis ball to car sized (64–4,000 mm)
Other–bedrock, hardpan, wood, etc.

Initials _____

* Peck, D.V., A.T. Herlihy, B.H. Hill, R.M. Hughes, P.R. Kaufmann, D.J. Klemm, J.M. Lazorchak, F.H. McCormick, S.A. Peterson, P.L. Ringold, T. Magee, and M.R. Cappaert. 2006. *Environmental Monitoring and Assessment Program—Surface Waters. Western Pilot Study: Field Operations Manual for Wadeable Streams.* EPA/620/R-06/003. US Environmental Protection Agency, Office of Research and Development, Washington, DC.

# Stream Benthic Macroinvertebrate
# Sampling Data Form Instructions[*]

**Name and ID:** Enter the name and ID number of the stream to be sampled. Provide both the USGS name (from topo map) and the local name (if applicable) that the stream is known by. Enter the name and ID of the sampling site on that stream that will be used for the normal sample. Use the same stream and site ID numbers as are used for normal water sampling at this same location.

**Location:** Enter the latitude, longitude, and datum of the X site (midpoint of the support reach).

**Sample Date and Time:** Indicate the date and time of beginning the biological sampling. Use 24-h (military time) format and indicate whether standard local time or daylight savings time.

**Field Team Leader:** Enter the name, affiliation, phone number, and e-mail address of the responsible field person.

**Water Sampling:** Check to indicate whether a water sample was collected for chemical analysis on the same day that this benthic invertebrate sample was collected.

**Site Documentation:** Check to make sure that a site documentation form has been prepared for this site; if not, prepare one.

**Number of Jars:** Indicate how many jars were needed to hold the sampled invertebrates.

**Replicate:** Indicate whether the sample at this site was replicated. If so, sample each transect for both the normal and the replicate sample before moving upstream to the next transect location. Assign a different unique sample ID number to the replicate sample. Enter the replicate site ID number.

**Number of Kick Net Samples:** Record the total number of kick net samples collected and combined to make up the normal sample.

**Substrate and Channel Types:** For transect A, check the appropriate substrate type (fine/sand, gravel, coarse, or other) and then check the appropriate channel type (pool, glide, riffle, or rapid). Then, do the same for transect B. Repeat until all transects are completed. See form for definition of substrate types and channel types.

**Comments:** Add any comments that may help to interpret and understand the stream biological and physical data. Describe the location of the support reach in sufficient detail to enable future field personnel to find this location again.

**Initials:** Initial the bottom of the form after carefully checking all entries.

---

[*] ibid.

# ZOOPLANKTON-SAMPLING DATA FORM

| | | |
|---|---|---|
| Site ID: _____ | Lake Name: _____ | Sample ID: _____ |

Tow Site Location: Latitude _ _ . _ _ _ _   _ _ _   _   Longitude: _ _ _ . _ _ _ _ _ _   Datum: _____

Site Type: ☐ Index ☐ Other (Describe): _____

Sample Date: _____

 Sample Time (24-hr): _ _ _ _   ☐ Standard ☐ Daylight Savings

Field Team Leader:   Name _____ Telephone _____

 Affiliation _____ Email _____

Water Sample for Chemical Analysis Collected on the Same Day as Zooplankton Sample? ☐ Yes ☐ No

Has a Lake Sampling Site Documentation Form Been Completed for This Site? ☐ Yes ☐ No If not, complete one now.

Net Opening Diameter: ☐ m ☐ ft ☐ cm ☐ in Coarse Mesh: _____ Fine Mesh: _____

Tow Type: ☐ Vertical ☐ Other (Describe): -

_____

Tow Length: _____ ☐ m ☐ ft

Mesh Size(s) Used (check all that apply): ☐ 80 μm ☐ 243 μm ☐ Other (Specify): _____

Number of Tows with Each Mesh Size:

    Coarse Mesh   ☐ One ☐ Two ☐ Other (Describe): _____

    Fine Mesh      ☐ One ☐ Two ☐ Other (Describe): _____

Number of Tow Sites Sampled on this Site Visit: ☐ One ☐ Two ☐ Other Specify: _____

Habitat(s) Sampled (check all that apply): ☐ Surface ☐ Pelagic ☐ Littoral

Number of Jars Used to Contain Sample:

    Coarse Mesh   ☐ One ☐ Two ☐ Three

    Fine Mesh      ☐ One ☐ Two ☐ Three

Sample Replicated: ☐ Yes ☐ No Replicate Sample ID: _____

Field Comments:

_____

_____

_____

_____

_____

_____

_____

# Lake-Zooplankton-Sampling Data Form Instructions

**Name and ID:** Enter the name and ID number of the stream to be sampled. Provide both the USGS name (from topo map) and the local name (if applicable) that the stream is known by. Enter the name and ID of the sampling site on that stream that will be used for the normal sample. Use the same stream and site ID numbers as are used for normal water sampling at this same location.

**Tow Site Location:** Enter the latitude, longitude, and datum of the zooplankton tow site location.

**Sample Date and Time:** Indicate the date and time of beginning the biological sampling. Use 24-h (military time) format and indicate whether standard local time or daylight savings time.

**Field Team Leader:** Enter the name, affiliation, phone number, and e-mail address of the responsible field person.

**Site Documentation:** Check to make sure that a site documentation form has been prepared for this site; if not, prepare one.

**Water Sampling:** Check to indicate whether a water sample was collected for chemical analysis on the same day that this zooplankton sample was collected.

**Net Opening Diameter:** Enter the opening diameter of both the coarse- and fine-mesh nets. Indicate the unit of measure.

**Tow Type:** Indicate type of tow.

**Tow Length:** Enter the length of water through which the nets were towed. Indicate the unit of measure.

**Mesh Size:** Enter the mesh size for each of the coarse- and fine-mesh nets.

**Number of Tows:** Enter the number of tows of the recorded length that were pooled to yield the sample for each net at the tow site. Most commonly, this will be one.

**Number of Tow Sites Sampled on This Site Visit:** Typically, zooplankton tows will only be collected from one lake site on a given lake visit, and this will generally be at the index site location. If other tow sites were sampled on this visit at other locations on this lake, record how many sites were sampled. *Fill out a separate Zooplankton-Sampling Data Form for each site location on the lake.*

**Habitat:** Enter the habitat sampled. Most commonly, this will be pelagic (open-water location).

**Number of Jars:** Enter the number of jars required to hold the sample for each of the net sizes.

**Replicate**: Indicate whether the sample was replicated. If so, assign a different sample ID to the normal and replicate samples. Enter the replicate sample ID number.

**Comments:** Add any comments that may help to interpret and understand the stream biological and physical data.

**Initials:** Initial the bottom of the form after carefully checking all entries.

# Index

Rapid Bioassessment Protocol (RBP);
*see* Rapid Bioassessment
Protocol (RBP)
Environmental replicate samples,
109–110
EPT Index (Ephemeroptera-Plecoptera-
Tricoptera), 210
Equipment cleaning protocols, 82–84,
85–86, 89, 186, 240
Eutrophication, 37
studies, 40–41, 43
surface water sensitivity to, 9

## F

Field blanks, 109
Filter blanks, 109
Fluoride (F), 40
Food chain, bioaccumulation of
contaminants in, 1
Forest Service (FS), 4
Fungicide deposition, effect of, 43

## G

Great Smoky Mountains, 38

## H

Hydrochloric acid (HCL), 84, 85
Hydrologic events, 20–21
water samples, effects on, 48, 149–150

## I

Index of Biotic Integrity (IBI), 179

## K

Kemmerer samples, 65
Kolmogorov-Smirnov test, 163

## L

Laboratory analyses. *See also* Chemical
analyses
biological samples, of, 199
bottle cleaning, 82–84, 85–86, 89

chemical analyses, 82
cooperation with field teams,
53, 54, 55
deionized water (DIW) rinses, 82, 83,
84, 85, 86, 89
filtration, 87, 89
overview, 81–82
preparation issues, 81
sample processing, 82, 86–87, 89–90
sample serial number (SSN), 86
sample storage, 87, 88–89, 89–90
Lake zooplankton. *See* Zooplankton, lake
Littoral zone surveys, 71

## M

Macroinvertebrates, stream. *See* Benthic
macroinvertebrates
Mercury (Hg), 1
atmospheric deposition, 5, 43–44
atmospheric transport, 10
bioaccumulation studies, 43, 44
cycling, 44
fish and aquatic organisms,
concentrations in, 5, 9–10, 44
methylated, 5
toxicity studies, 43, 44
Method detection limits (MDLs), 95–97,
137–138
Model of Acidification of Groundwater in
Catchments (MAGIC), 20, 21
Most probable value (MPV), 112
Multimetric indices (IBI), 201–202
usage, 203

## N

National Environmental Laboratory
Accreditation Conference
(NELAC), 117, 118
National Laboratory Accreditation
Program (NELAP), 118
National Lakes Survey, 193
National Surface Water Surveys (NSWS),
72, 113, 139
National Wadeable Streams Assessment
(WSA), 203